"I'm so grateful to internal family systems (IFS) trainer Sand C. Chang for writing this book. It so skillfully applies my life's work to the queer and trans community, a group that, traditionally and particularly lately, contains some of the more persecuted exiles in our culture. I not only highly recommend it as a potent workbook for that community, but also for anyone who wants to better understand and/or help them."

—**Richard Schwartz, PhD**, developer of the IFS model

"I love this workbook! It's a gentle, accessible, and nuanced guide to working with one's internal parts, specifically for queer and trans people. Sand C. Chang encourages the reader to welcome *all* parts—making room for the tender ones inside that may have been marginalized, dismissed, or denigrated by the heteronormative world. What a beautiful gift to us and to the field of IFS."

—**Colleen West, LMFT**, IFS consultant, and author of *We All Have Parts* and *The IFS Flip Chart*

"Without exaggeration, this workbook is immediately indispensable! Sand C. Chang gently guides queer and trans readers into a calm and compassionate IFS practice space. Their personal illustrations, captivating reflection questions, innovative worksheets, and courageous meditations amplify our inner wisdom. Chang invites us into embracing our unique inner ecosystems while centering collective care. This book is a bright star illuminating how we may become more skillful at radical belonging."

—**Ash(ley) Gregory, LMFT**, (she/her), queer psychotherapist and proud Queer and Trans Internal Family System (QTIFS) teaching assistant who transforms shamefulness into spaciousness with curiosity, playfulness, and hope

"Sand C. Chang's knowledge of IFS, commitment to body liberation, and fierce love of community is felt throughout this book. The meditations, reflections, and guided prompts allow one to feel Chang's support as they invite the reader to get curious about and deepen their relationship to themselves. *All Parts Welcome* is a gift to queer and trans people interested in IFS as a healing practice."

—**Nic Wildes, LMHC**, certified IFS therapist, IFSI-approved consultant, and cofounder of QTIFS

"Sand C. Chang lovingly guides the reader through introducing IFS theory to how this work intersects with our bodies, gender, and sexual orientation. On this journey I could feel a settling in my body, an internal understanding that as a queer and trans person this resource really is for us—a tool that connects our inner world with the external challenges of heterosexism and cis-supremacy while providing a blueprint for healing."

— **Damon Constantinides, PhD**, sex therapist in private practice, CIIS Sex Therapy certificate instructor, and coauthor of *Sex Therapy with Erotically Marginalized Clients*

"*All Parts Welcome* is an essential resource for LGBTQIA2S+ individuals seeking healing, and for therapists supporting them. Sand C. Chang expertly blends IFS therapy with the lived experiences of queer and trans people. This book empowers readers to understand their inner world, address the impact of societal biases, and reclaim self-led self-care and community. It's such a powerful tool for fostering inclusive and supportive IFS practices!"

— **Natalie Y. Gutiérrez, LMFT**, author of *The Pain We Carry* and *The Pain We Carry Workbook*

"With *All Parts Welcome*, Sand C. Chang has created a model we can trust—one that is inherently affirming, deeply compassionate, and fully aligned with the lived experiences of queer and trans people. This workbook seamlessly integrates IFS in a way that feels both accessible and transformative, offering a much-needed resource for healing, self-understanding, and coming home to yourself."

— **Rebecca Minor, LICSW**, gender specialist, consultant, and author of *Raising Trans Kids*

All Parts Welcome

The Queer and Trans Internal Family Systems Workbook

Sand C. Chang, PhD

New Harbinger Publications, Inc.

Publisher's Note

This publication is designed to provide accurate and authoritative information in regard to the subject matter covered. It is sold with the understanding that the publisher is not engaged in rendering psychological, financial, legal, or other professional services. If expert assistance or counseling is needed, the services of a competent professional should be sought.

NEW HARBINGER PUBLICATIONS is a registered trademark of New Harbinger Publications, Inc.

New Harbinger Publications is an employee-owned company.

Copyright © 2025 by Sand C. Chang
New Harbinger Publications, Inc.
5720 Shattuck Avenue
Oakland, CA 94609
www.newharbinger.com

All Rights Reserved

Cover design by Sara Christian

Acquired by Elizabeth Hollis Hansen

Edited by Karen Levy

Library of Congress Cataloging-in-Publication Data on file

Printed in the United States of America

27   26   25

10   9   8   7   6   5   4   3   2   1   First Printing

To all the queer and trans people who persist in self-determination, love fiercely, make and celebrate art, fight for liberation, and show up again and again for each other, and to queer ancestors and trans-cestors, who fortify us with their gifts.

Contents

	Foreword	vii
	Introduction	1
1	What Is Internal Family Systems?	5
2	Principles and Goals of IFS	17
3	The You-est You: Tapping into Self Energy	33
4	Protector Parts	51
5	Unblending While Respecting the Protective System	67
6	Exiled Parts	81
7	How Parts Heal	87
8	Coming into Our Genders	103
9	Sexuality and Relationships	115
10	Relating to Our Bodies	131
11	Recognizing Our Strengths	145
12	Persisting in Self-Determination: Navigating the External World	151
13	Mental Health Support	161
14	Establishing a Continuous Practice	167
	Acknowledgments	171
	Glossary	173
	References and Further Reading	177

Foreword

It is both my pleasure and honor to write the foreword to this leading-edge offering of healing for members of queer and trans communities. As the first Black Internal Family Systems (IFS) Therapy trainer, it is my mission to assist other professionals in sharing the transformational potential of IFS with marginalized groups, and I celebrate what Dr. Sand Chang has created with this work. I came to know of Sand through a mutual friend in the IFS community who shared Sand's groundbreaking course, *IFS Foundations for Trans, Nonbinary, and Queer Clinicians*. Knowing their work made me eager to meet them. We met when Sand joined my IFS consultation group for therapists and practitioners. I was instantly impressed by the ease with which they shared the potential impact of IFS on the communities they serve. Sand had an uncanny understanding of the complexities of IFS. They expressed a desire to combine their knowledge of the needs of this community with the richness of IFS and its capacity to unburden queer and trans folks.

While traveling in Costa Rica recently, I overheard a conversation between two people from the United States. A man was expressing (quite vehemently) his disdain for parents who accept their nonbinary children instead of demanding that they choose a gender. He went on and on criticizing the parents and saying nonbinary people are confused. I thought, "I am in the midst of writing the foreword to a book that helps queer and trans people develop a relationship with parts of themselves who may have been harmed by someone with similar views." In that moment, I felt a wellspring of deep compassion for the experiences of this community and appreciation for Sand for this compelling book.

Sand has beautifully adapted the concepts of IFS to fit the unique experiences of queer and trans folks instead of insisting they contort to fit the model. This is a true skill. As I read this book, I felt Sand's care poured onto each page. They have a way with language that connects the reader to the message. The concepts of IFS are explained in ways that will benefit folks whether they are familiar with IFS or not. The examples thoroughly support the learning. Sand skillfully describes the complexity of the queer and trans system and the numerous ways they are pressured into assimilation. This book nicely lays out a path for profound, deep emotional work.

I can't think of anyone better qualified to bring this healing work forth. This book is a collection of Sand's vast knowledge of IFS, their experience working with queer, nonbinary, and trans people and the therapists who serve them. I highly recommend this book to members of this community who want to engage in a transformational journey with support from someone who knows your challenges firsthand. Thank you, Sand, for writing this much-needed book for self-discovery and healing.

—Tamala Floyd, LCSW, IFS Solo Lead Trainer and author of
Listening When Parts Speak: A Practical Guide to Healing with Internal Family Systems Therapy and Ancestor Wisdom

Introduction

"We aren't one-dimensional. Yes, we can be multiple people at once, and we can still be authentic in our identities. Don't hold yourself to only one version, one story. Our layers are what make us unique; within the textures and contrasts are what makes us magnetic."

—*Anthony Perrotta*

This is a book about connection. As humans, we're relational creatures. We seek contact, comradery, closeness, and belonging with other humans. We spend time getting to know other people and deciding how we want to relate to them. Some of us might take for granted the time and effort we spend focusing on our relationships with the people, places, and things outside of us. This book is about taking all that energy and intention that we put *out there* and redirecting it toward our own internal ecosystems. We face the task of building secure and loving bonds with/in ourselves, and, in turn, this starts to transform and cause ripple effects in how we relate to our external circumstances. This is about healing from the inside out.

This book is written specifically for you, the queer and trans reader, someone seeking a different way to grow and heal through Internal Family Systems. In the IFS world we often say "all parts welcome" to express an openness and welcoming to our nuanced, complex, and sometimes contradictory ways of being. As queer and trans people, we may not always feel like it's okay to express our full selves. We typically don't expect all of our parts to be welcomed because they haven't been. It's important to acknowledge this reality—that it's not just in our heads, that there are real reasons why we (consciously or unconsciously) hide parts of us away from the world. This is particularly salient at this moment in time, a time when our rights as queer and trans people are being heavily attacked and debated, whether it be in education, sports, family building and parental rights, the right to medically necessary health care, access to restrooms, accessing the same kinds of benefits that cisgender and heterosexual people take for granted, or our freedom to wear whatever the hell we choose to feel like our most fabulous, authentic selves. The list goes on, and it takes a toll on our psyches. Our fear and pain are real and valid.

I can't promise that IFS will solve these problems—in all of these cases, systems change is necessary. But what I can offer is the possibility that doing whatever we can to know and heal our parts will give us strength to keep going and to show up for ourselves and each other in the face of adversity. Even in situations when we can't show our full selves to the world, we may find some comfort in trusting ourselves and knowing our own stories. I can also offer years of experience witnessing my clients heal. To watch and feel the difference as guarded parts are able to relax back, knowing they can trust in the client's capacity to lead, is beautiful and powerful.

Situating Myself

I love queer and trans people. I fucking love us, I really do. That being said, I don't in any way speak for or represent all queer and trans people. Like many humans, I've lived a mix of privileged and marginalized social locations that influence my perspectives as well what's outside my awareness. Specifically, these social locations include: being Chinese American, a child of immigrants, a US citizen, neurodivergent, queer, trans, nonbinary, genderfluid, X-ennial, upper middle class, midsize, spiritual, in recovery, and currently living on unceded Tongva and Ohlone land.

I'm a clinical psychologist, Body Trust provider, DEI consultant, trainer, astrologer, and lifelong student. Much of my career has been dedicated to body liberation, particularly working with/in trans health and eating disorders. I envision a world in which queer and trans people don't have to prove anything to be seen as valid; a world in which we can just go about our lives without fear of being attacked, erased, or scrutinized; a world in which all parts of us are welcome, especially by us.

My IFS Journey

I was in my postdoctoral psychology fellowship when I first learned about IFS through a colleague's very enthusiastic presentation at our monthly staff training. At the time, part of me was curious about this approach. It sounded interesting and slightly poetic. But that curiosity was quickly squashed by a much louder part of me that thought, *Managers? Firefighters? This is cheesy.* I was turned off by the jargon and what seemed to be too "out there" for me. But in this training I was also introduced to a heart meditation, which, despite my inner eye roll, sparked a deep resonance within me. Protectors around my heart? I could relate to that! I immediately purchased some recorded IFS meditations by Richard Schwartz, and I listened to that heart meditation almost every day for years. There was something to this, but my system wouldn't allow me to explore it further.

Fast-forward to ten years later, when I was fairly established in my career as a therapist. I was in a longstanding EMDR consultation group with Colleen West, and slowly, one by one, every single person in the group became an IFS therapist. I started to get lost in our case consultations, and of course, then a part of me (that hates not knowing what's going on) rushed to get into an IFS training. This was still before IFS became wildly popular in the general public, so I lucked out and got into a small training in Santa Cruz. Little did I how much was about to change.

To provide some context, I've identified as a "patient" from a young age. Early on I developed a variety pack of mental health "symptoms" and "syndromes" that I believed defined me. Luckily, I was such a mess that by the time I was a young adult I'd landed in twice-weekly therapy and a slew of recovery programs. I started to put together a full life, learning how to build relationships and take care of myself. I became a therapist and continued to engage in different kinds of therapy. All of these things were helpful in some way, but what I didn't realize was that much of that healing work was motivated by the (well-intentioned) part of me that wanted to be seen as good, healthy, spiritual, and respectable. My growth work allowed me to "function," but young parts of me were still wounded. Years of talk therapy instilled an aptitude for narrating *for* my parts, telling their stories without ever listening to them directly. I knew *about* myself, but I didn't truly *know* myself.

My own IFS therapy opened me up to a way of being that I didn't know was possible, and that's why I'm pretty much in love with it. Is it a perfect model? No, nothing is! But it's the approach that has allowed me to be at peace with myself the majority of the time. I know my parts well: the critics, the speedy and impulsive ones, the hyperverbal ones, the ones who shut down, the ones who want to attack and defend. And we're cool. They don't really need to jump in like they used to because they trust me. I've also gotten to reclaim and deepen my relationships with parts of me that are sensitive, carefree, creative, silly, pun-obsessed, and full of wonder. And there are parts of me that I'm yet to meet. It keeps things fresh!

Most of my time is now dedicated to IFS. As a trainer with the IFS Institute, I get to learn from people I admire and who are willing to be in conversation about how the model is evolving. Teaching IFS to queer and trans providers with my colleague Nic Wildes is truly a highlight of my work. And I get to play! I love finding new and different ways to integrate parts work with eating disorders, 12-step recovery, astrology, and more. Healing work doesn't have to be serious all the time; I encourage anyone who's engaging in IFS to find ways to get creative and playful with it.

How to Use This Book

There's no right way to use this book, but I do suggest really making an effort to use it. Give yourself the opportunity to benefit from the practice of IFS and parts work by reading and doing the exercises in the entire book. When you feel stalled, you can use that as an opportunity to get curious about and connect with any parts of you that might have concerns.

Each chapter offers some information about IFS and some exercises and meditations that will help you spend time with your internal system. You can find copies of the exercises and audio recordings of the meditations at the website for this book: http://www.newharbinger.com/55282. Some of the exercises may resonate more than others for you. If that's the case, keep using the ones that you like but also come back to the other ones at some point, as your experience of them may change with more experience. Chapters 1 through 7 introduce you to the process of working with parts, and chapters 8 through 12 explore topics that are geared specifically toward queer and trans communities.

If you're working with a therapist, you may want to work through this workbook with them. Know that this book can help you get equipped to do IFS work, but the real gems of this work will come through with time and practice. Chapters 13 and 14 offers tips for how to get mental health support and establish a continuous practice.

A word about parts and pronouns: For people of all genders, **we don't assume the gender or pronouns of any part unless or until the part expresses itself in a particular way.** We can use the pronoun "it" when we're referring to the *part* (not a person) and don't know much about it yet, but we can shift if the part indicates something else.

Without further ado, let's get started!

Chapter 1

What Is Internal Family Systems?

Internal Family Systems (IFS) is an evidence-based healing approach developed by Richard Schwartz, PhD, in the 1980s while working with clients with eating disorders. Schwartz noticed that many clients described having parts of themselves with conflicting feelings or perspectives. Being a family therapist with an understanding that clients needed to be viewed within the larger context of their families, communities, and environments, he realized that this could also be true for what's happening *within* a person's internal ecosystem. The parts inside us could be understood as a sort of inner family. Like other families, these families consist of members with different personalities, hopes, motivations, and roles. When there's tension within the inner family, that can negatively affect how a person experiences themself, others, and the world. When there's harmony within the inner family, there's more ease and flow in a person's life.

One of the most important features of IFS is the belief that **we all have a Self** (alternately referred to as **Self energy**): an innate wisdom, knowing, or consciousness that has an abundant capacity to witness and offer a healing experience to every part of us. In IFS communities, we often say "all parts welcome" or that there are "no bad parts" (Schwartz and Morissette, 2021). This is very different from treatment approaches that characterize parts of us (sometimes expressed as symptoms such as depression or anxiety) as having bad intentions or being "out to get us." This is what drew me the most to IFS: the recognition of parts that help us cope and *survive* even if their strategies or behaviors aren't always helpful in the long term. It's an incredibly compassionate view of what exists within the human psyche.

The concept of **multiplicity** is central to IFS. This explains why, as human beings, we aren't always the same day to day or even moment to moment. We aren't always consistent with ourselves. We may feel one way and feel the opposite way at the same time. When we're indecisive, it's often because we have parts that aren't in agreement. The concept of multiplicity can be incredibly freeing: it can release the expectation that every experience (feeling, sensation, belief, action) we have must assimilate to a singular way of being or relating to ourselves and the world.

Recognizing multiplicity as normal (not in any way bad, wrong, or sick) allows us to embrace the complexities of who we are. You probably don't have to think long to come up with an example of how different situations might pull for different ways of reacting to a person, place, or event. For example, in some situations you may feel more extroverted, while in others you may feel more introverted. You could consider this being ambiverted. Another framing is that some parts of you are more extroverted and other parts are more introverted, and different situations call for different parts to be expressed.

When we have a strong pull to express ourselves in a particular way, it might be because we're **blended** with a part. Sometimes we're so blended that we can't access other parts of us. In IFS, we aim to build more capacity to lead with Self energy so that we can more easily unblend and get to know different parts of us.

Pause and reflect: What's your reaction to the idea of multiplicity? How do you relate to the idea that there are different parts of you that show up in different situations?

_____

_____

_____

_____

Each part of us has different qualities and energies. Some might feel urgent or intense, while others may feel slow or calm. Some show up as body sensations, while others might appear like thoughts. An anxious part that's obsessively trying to make a decision by making a lot of pros and cons lists might be experienced more in the thought realm, while another anxious part that's terrified of public speaking might be expressed more through physical sensations (e.g., racing heart, shallow breathing, sweaty palms).

IFS suggests that we have three different categories of parts: managers and firefighters (collectively referred to as **protectors** or **protector parts**) and exiles (parts that are being protected by the other two). **Managers** or **manager parts** are proactive parts of us that operate our day-to-day functions. For example, before making a large online purchase, a manager part might prompt you to check your bank account or make sure the item is needed. **Firefighters** or **firefighter parts** are exactly what they sound like: they fight fires, meaning they jump in to help or fix when something seems urgent to them. They're more reactive than proactive. In the example of making a large purchase, a firefighter might impulsively click the "purchase your order" button if it feels very strongly that you *must* have the item, regardless of the situational factors that may or may not support the purchase. These two parts are in a **polarization** with each other and can create internal pressure that we may not even be aware of! Relatable?

The third type of part is referred to as an **exile**. This is because it's typically outside of our conscious awareness. Very often exiles are kept out of sight by our protector (manager or firefighter) parts. The thing is, it's the exiles that need attention and healing the most. Exiles are often young parts, sometimes parts that got "stuck" emotionally or developmentally in a particularly challenging moment in the past. They might try to get our attention, sometimes expressed through the nervous system or body. Exiles carry wounds or painful beliefs about themselves or the world. We refer to these wounds or beliefs as **burdens**.

In IFS, we have a way of helping our parts release the burdens that they've been carrying. For some people, doing this kind of parts work is similar to what's considered "inner child" work. But if someone asked me about my inner child, I'd ask, "Which one?" There may be many young parts that need attention. And, because every person's internal system is different, you may not experience your more vulnerable or exiled parts as children. You might experience them as animals, body sensations, or simply energies.

All of our parts are born out of experience, both positive and negative. As I'm writing this, I notice a part of me that resists making blanket statements or calling any of our experiences "negative." That being said, some of our experiences have been more uncomfortable or painful than others or perhaps had a negative impact on our overall well-being. And some of our parts might work harder than others to keep us safe or allow us to function[i] in our everyday lives. Having parts is just "part" of being human.

---

i   The term "function" is loaded and often used or interpreted in an ableist way. When I use the word "function," it's with the awareness that we all live in a world that rewards those who function or perform based on mainstream, dominant culture standards and often requires us to strive to meet these standards for survival.

Throughout this book you'll get more acquainted with the different parts in your unique system. For now, just know that labeling parts or knowing what kind they are isn't always necessary. Just like when you meet a new person, it's helpful to just listen and get to know the person before jumping to any conclusions about what "type" of person they are. If you notice yourself getting caught up in trying to figure out what kind of part you're working with, that may be a part that's overthinking instead of simply being open to getting to know the part. As much as you are able to in the moment, try to approach each part with openness, curiosity, and presence. And if that's not available, that's okay. You can work with whatever is coming up in response to your parts.

Pause and reflect: How do you feel about the idea of getting to know parts of you better and building a caring relationship with them? Notice any feelings that come up if you even think about trying to connect with parts. Notice any protectors that might show up and jot down their concerns.

_____

_____

_____

Following Trailheads

In IFS we use the term "trailhead" to refer to a starting point or an entry into an issue that's raising concern for us. In response to a therapist asking, "What do you want to focus on today?" or a friend saying, "It seems like something's bothering you. What's going on?" we might say whatever is troubling us. After identifying a trailhead, we can then follow a path that will help us go deeper and identify parts/protectors that are active in whatever issue we're focusing on. In some of the chapters in this book, you'll follow a trailhead as far as you can go with it (e.g., working with protectors, then helping to heal exiles). In other chapters you'll just be identifying trailheads related to certain themes or topics in your life. If you don't know which trailhead to start with, just start going down the path. You may find that two seemingly separate issues (i.e., different trailheads) lead to the same place.

The Path to Healing: Communicating with Our Parts

So far in discussing this model, we've been talking *about* parts. There are other kinds of therapy or healing work that might frame us as having parts. Even in casual conversation people talk about having parts that feel one way and parts that feel another. So how is the IFS approach different?

In IFS, building *relationship with* our parts (often referred to as **Self-to-part relationship**) is the way that we heal. That means we direct our attention toward our parts and actually communicate directly with them. We can talk about our parts all day and create elaborate hypotheses *about* them, but that doesn't mean we have a true connection *with* them. If we meet a new person, ideally we don't just look at them, narrate a story about them, and then assume we understand them. We'd want to ask the person some version of "Do you want to have a conversation?" or "Tell me about yourself!" The same goes for how we relate to parts. We invite them to speak for themselves as we *listen*. Doing so, we learn new information that we could not have come up with ourselves. Don't worry too much about *how* to communicate with your parts right now. Just know that this internal dialogue is key to the process.

You may notice a part that's really loud and requires no introduction when it appears. There are other parts that are quieter or that you may not even know exist. Therefore, some parts will be easier to hear from or interact with than others. Like any relationship, connection and understanding develop over time or with repeated contact. Some of these relationships with your parts will take time, possibly more time than other parts might prefer! Over time, you'll be able to have more open conversations with these different parts of you, which you might consider your inner family. The following worksheet will help you identify which parts might be having feelings or reactions in this moment. You may want to write down the first things that come to your mind, then take a break and come back to write more later. That might make it easier to gain a fuller perspective from more parts. A copy of this worksheet can be found at http://www.newharbinger.com/55282.

Worksheet: Recognizing Parts

What's your reaction to the idea of welcoming all parts of you? Notice any parts that have reactions or objections to the idea and write down any concerns they have.

_____

_____

_____

_____

Think about the last big decision you made. How much did you try to consider or listen to conflicting feelings or perspectives within you?

_____

_____

_____

_____

Reflect on your day so far. Which parts have shown up to respond to or take care of different situations? Do you feel that managers have been more present, that firefighters have been more present, or both?

_____

_____

_____

_____

To start getting to know a part of you, we're going to explore a relatable everyday situation. You may use the questions in this meditation to build your practice of getting to know any parts that you're aware of. An audio recording of this meditation can be found at http://www.newharbinger.com/55282.

Meditation: A Household Chore

1. Find a relatively comfortable position and start to notice your breath. Notice what it feels like with each inhale and exhale. Take as much time as you need to get situated.

2. When you're ready, call into your mind a household chore that you do on a regular basis, for example, washing dishes, doing laundry, or making your bed. Notice any reactions you have at the thought of this task. Just notice. At this moment you don't need to do anything, just notice however your reaction shows up.

3. Notice any emotions that arise and how you're experiencing them.

4. Notice any body sensations.

5. Notice if any images come to mind.

6. Take note of any thoughts, beliefs, or messages. You may encounter parts of you that want you to get this chore done, parts that resist or just don't want to do it, or parts that distract. You may experience a polarization or sense of tension between different parts of you. As much as is possible, just notice, not having to make any decisions or take any sides.

7. If you do have strong opinions or feelings toward these parts, such as frustration or judgment, ask those opinions or feelings to move to the side so that you can remain open and present. If that's not possible, just stay right there and notice what's coming up. Whatever you're noticing, try to let it know that you're aware of it, that you're listening. You might even say to a part of you, "I hear you. That makes sense." You might also thank the part by saying, "Thank you for letting me know."

8. Communication with your parts may be verbal, but not always. Sometimes you might imagine sending a feeling or an energy, such as a sense of care. And then notice what happens. Are there any shifts? Any new sensations, feelings, images, or thoughts? Any softening or quieting or settling? If nothing happens, that's okay. This is just about trying on a practice, starting a conversation. For now, we're just making contact, and at a later time you'll be able to have a more in-depth conversation.

9. Take the time you need to listen in to different parts of you, not having to do anything, but just with the intention to learn and understand them better.

10. When you're ready, you can thank all parts of you that allowed you to notice them. If you'd like, you can take some time to journal about what this meditation was like for you and which parts showed up.

Pause and reflect: What came up for you during the meditation?

_____
_____
_____

Why IFS for Queer and Trans Communities?

As queer and trans people, many of us understand what it's like to have to hide or not fully acknowledge parts of us. And this is understandable: many of us have internalized negative beliefs about what it means to be lesbian, gay, bisexual, queer, trans, nonbinary, or basically anything that deviates from dominant culture expectations for gender or sexual expression. Many of us have experienced criticism or rejection from people or systems that can't embrace the fullness and beauty of who we are. We may have become adept at blending in or shape-shifting to fit in, maintain connection with others, or avoid loss or danger. When we can honor that these ways of coping are parts that are working hard to keep us safe, there's hope of building a trusting relationship with them.

IFS, at its core, is about inclusion and belonging. IFS is a model that can hold the complexity and paradox of all our identities, including challenging experiences and joyful ones. Rather than simply focus on our queer or trans identities, there's space for us to invite in all the ways that those identities intersect with other important facets of who we are: race, ethnicity, religion, class, and ability, to name a few. Honoring our multiplicity allows us to respect the different experiences that have shaped us, as well as how these experiences play off each other. It can be immensely comforting and a relief to know that with Self energy, every part can be heard and appreciated.

Beyond whether and how we're received in all our complexities by our social environments, we have our own relationships to our gender identity, gender expression, sexuality, and how we build relationships. For many queer and trans people, realizing that the expectations or ideals prescribed to us didn't fully encapsulate who we are (or perhaps not at all!) may have created an opening to experiencing ourselves and our identities in much vaster ways. It may have allowed us to think more expansively about options and possibilities for how we relate to gender or sexuality. **Our capacity to honor our truths even when the external world refuses to is a reflection of our own Self energy.**

As queer and trans people, one word often isn't enough to describe different aspects of ourselves, so we choose more than one. Or perhaps we feel that some parts are more present on some days than others. For example, someone who experiences their gender as fluid, sometimes more masculine and sometimes more feminine[ii] and sometimes neither, might have their gender expression be influenced by what's happening in their internal system. Perhaps on some days a more masculine part may want to be seen or expressed more, while on other days a more feminine part may want to be expressed more, while on other days there's an expression of both at the same time. These varying expressions might give permission for parts with different gender identities to be expressed. This is just one example. We'll spend more time exploring the ways that your parts influence your gender and sexuality in chapters 8 and 9. In the meantime, the following worksheet will help you reflect on how multiplicity influences your cultural identities, whether it be gender, sexuality, or other significant cultural factors in your life. You're invited to notice the reactions your parts have to different aspects of identity, as they may differ considerably. A copy of this worksheet can be found at http://www.newharbinger.com/55282.

---

ii   The concepts of masculinity and femininity are highly contextual and culturally dependent. What is considered masculine in one culture or at one point in time might be very different from what is considered masculine in a different culture or at a different point in time. When I use these words, I do so with the understanding that each person gets to decide what these terms or concepts mean to them and to what extent they subscribe to systems that impose certain rules for masculinity and femininity. Further, many of the assumptions that are made about masculinity and femininity are embedded in Western or colonial frameworks.

Worksheet: Multiplicity and Your Cultural Identities

When you've experienced any sense of disconnection or loss due to your queer and/or trans identity or others' reactions to it, how have you coped? Which parts may have helped you survive at different points in your life?

_____

_____

_____

When you've experienced support or connection in relation to your queer and/or trans identity or others' reactions to it, what has that been like? Which parts may have reacted or responded to that?

_____

_____

_____

When you consider the many different facets of who you are, including your cultural identities, how have you adapted through different circumstances that may or may not have supported those identities? For example, consider your most recent workplace (or any other environment you spend a lot of time in). Do you feel like you can be your whole self? Which identities feel seen? Which identities are more hidden?

_____

_____

_____

_____

Which aspects of your cultural identity or identities do you know well? Circle these. Which aspects of your cultural identity or identities would you like to get to know better? Underline these. Here are some examples, and you can also write in your own.

- Racial identity
- Ethnic identity
- National origin
- Sexual orientation
- Gender and/or gender identity
- Religion and/or spirituality
- Socioeconomic class
- Immigration history or lineage
- Ability or disability status
- Body size
- Neurotype or neurodiversity
- Other: _____
- Other: _____

If you have a relationship to any particular queer and/or trans subcommunities, which parts of you feel most comfortable in them? Which parts aren't as comfortable being seen in them?

_____
_____
_____
_____
_____

Summing It Up

In this chapter, you were introduced to the basic components of IFS, including multiplicity, welcoming all parts, the three kinds of parts, and the importance of the Self-to-part relationship. You read about why IFS is such a good fit for supporting queer and trans communities. In the next chapter, you'll learn more about the underlying principles and goals of IFS and have an opportunity to identify your own goals.

Chapter 2

Principles and Goals of IFS

Core Principles of IFS

As queer and trans people, we know all too well what it's like to be taught certain rules for how to "be" in the world, and often these rules don't allow us to express ourselves authentically. The ways that we're taught to view or judge ourselves is usually not conducive to our healing and liberation. Like any approach to self-understanding, IFS is based on some core principles. While these concepts may sound simple, they differ from what the dominant culture has taught us. As you read, consider how IFS principles align (or don't align) with your worldview or what you've been taught. If you have parts that feel skeptical about IFS, those parts are welcome too! If you have parts that feel eager or even impatient to start working with your system, I encourage you to read through this chapter for any parts of you who may need a little more information.

Principle 1: We all have a Self or Self energy.

We all have Self energy—the capacity to be present, wise, and loving toward ourselves and others. As humans we're born with this capacity. Though we all have Self energy, it might not always feel like we do. It's not that some people have more Self energy than others, but rather that certain conditions allow each of us to *access* more or less of our Self energy. Some days you might be more resourced than others. This might be because you got enough sleep, spent time with a loved one, or are wearing soft clothes (yes, sensory needs really do make a difference!).

Okay, so then what does this mean? Does this mean we should just love everyone around us? Bypass our feelings by invoking toxic positivity? Nope, that's not where we're going with this. When people act in ways that are unkind or harmful, it's hard to recognize the good in them. Protective parts of us can quickly categorize others as safe or unsafe, sometimes for our survival. The point here isn't to ignore the parts that tell us we need to distance ourselves from potential harm, but to broaden our view by acknowledging that sometimes there are complicated reasons for why we all act the ways that we do.

Say you have a family member who makes a lot of demands on you for attention, time, or labor. On a good day, you might have more patience or feel able to set boundaries. On another day, when you have the flu or are experiencing a lot of pain, you might be more irritable or reactive. It might be harder to access a sense of calm or clarity. In other words, there might be more blocks to accessing Self energy. In each of these examples you could fall into the trap of labeling yourself as patient, or clear, or irritable, or reactive, as if these traits define you. You may have parts that express themselves as these traits, but these traits don't define you or negate the fact that you have something inside or throughout your being that's more compassionate and can deal more effectively with challenges as they arise. When we have more access to Self energy, we're more **Self-led**. When we don't have as much access to Self energy, we tend to be more blended with parts or **parts-led**.

Self energy is like the sun. We can trust it exists even when we can't see it, even when it's cloudy or stormy or eclipsed by the moon. Sometimes people say that the sun has "come out" when it's actually been there all along. In IFS, we address the "weather" that might block us from basking in our own sunlight.

Pause and reflect: People aren't all good or all bad, but sometimes we/they have more or less access to Self energy. What are your reactions to this idea?

_____

_____

_____

What conditions allow you to feel most connected to your Self energy? Consider personal or environmental factors, the people you spend time with, or activities that you engage in.

_____

_____

_____

Principle 2: Multiplicity is universal to being human.

In addition to having Self energy, we all have parts. Multiplicity is just an aspect of being human. From an early age, we may have been taught that there's an expectation for how we're supposed to be. Sometimes that expectation is based on our familial or ethnic background, our (assumed) gender, our religious background, and the list goes on. These external pressures are felt on the inside, and we learn to hide away parts of us that seem in conflict either with others' expectations or with the more socially acceptable parts of us.

In addition to this socialization, some schools of thought in the mental health field consider the idea of "parts" to be unhealthy; these approaches might aim to do away with parts or have them dissolve into one cohesive whole. While at times we might not like all of our parts, they exist for a reason. There's nothing inherently bad or sick about having parts, even ones that manifest as mental health symptoms. Accepting that we have parts allows us to be with who we really are, not just the aspects of us that are deemed acceptable by others.

It's not just "normal" to have parts but it's also something to be honored and celebrated. Multiplicity is often what makes us the beautiful and complex beings that we are. As queer and trans people, we know that things aren't always what they seem or are expected to be; the things that people don't see can be the best parts of us!

Pause and reflect: How much or how little have you allowed yourself to embrace your multiplicity? Do you have a sense as to why this might be the case?

_____

_____

_____

When you were growing up, was multiplicity allowed? If so, where was it allowed? Where was it discouraged?

_____

_____

How easy or difficult is it for you to consider that other people have multiplicity?

_____

_____

Principle 3: Parts adopt roles to help us cope with life.

Some parts we're born with, while there are others that come into existence because of our life experiences. From a young age, even before we know who we are, we face the task of adapting to the messages and rules of our family or the culture around us.

For example, if we're praised for knowing the right answer to a question, a part might work hard to make sure it continues to get the right answers to other questions. This isn't necessarily a bad thing, but if taken to an extreme this can lead to needing to be "right" all the time. When we do something that others disapprove of, a part may use the strategy of making sure we don't do that thing again. For example, if we were yelled at as soon as we expressed disagreement with an early caregiver, perhaps a part learned to be overly agreeable or compliant to avoid further punishment.

Our protective parts take on certain roles to either keep us safe from people in our external environments *or* keep us safe from feeling overwhelmed by other parts in our internal systems. Some parts might like their roles, while others might feel stuck and not know that there could be other possibilities for them. Sometimes our parts actively try to keep other parts from being seen.

The process of hiding our parts or our multiplicity is a form of **masking**, or the process by which (consciously or unconsciously) we behave in ways that aren't true to ourselves in order to be safe or fit in. Masking is often discussed in the context of neurodivergence, but it applies broadly to anything we do to suppress our authentic selves in service of the status quo. A related concept is **code-switching**, which is the skill of presenting ourselves or behaving differently depending on the social context. For example, we may have internalized an idea of "professionalism" (the most effective way to show up at work) that isn't in line with who we really are; this may lead us to suppress our personality so that we can blend in with dominant culture identities (e.g., white, non-disabled, cisgender, binary gender, neurotypical, etc.). Sometimes code-switching is intentional, strategic, or even for survival; therefore, it's important to name that it's not inherently bad. But it takes work and sometimes a toll on our emotional well-being! Masking or code-switching aren't deceptive; they're ways that our parts protect us and help us adapt to external pressures.

Pause and reflect: What's your reaction to the idea that our parts take on roles to protect us or help us cope?

_____

_____

_____

How do you relate to the concepts of masking and/or code-switching? How have they played a role in your life as a queer and/or trans person? Or as a person with any number of personal or cultural identities in your social environments?

_____

_____

_____

Principle 4: All parts have a positive intention or function.

Even when their behavior impacts us negatively, every part of us has a positive intention. This might be hard to believe, because sometimes we do things that aren't in our best interest or are harmful to ourselves or others.

For example, a person with severe social anxiety might have a part that tries to help them by using alcohol to numb painful feelings. Unfortunately, overly relying on substances can have negative or dangerous consequences, but the intention of the part who engages in drinking is positive (i.e., to protect). In IFS, we actually try to *help* protectors who might be causing us problems, which may mean offering them the possibility of a different *strategy*, one that doesn't have harmful consequences.

Even though our parts have positive intentions *for us*, we're still responsible for addressing their actions when they negatively impact us or other people. We can't bypass accountability by saying "It was just a part!" On the flip side, we don't want to condone others' unacceptable behavior by saying, "It's just a part of them." We can maintain an understanding that our parts have positive intentions for us, other people's parts have positive intentions for them, and we still need to own the impact of our actions and try to engage in repair when needed.

Pause and reflect: Think of a time when a part tried to help you but ended up causing a problem. How is it for you to try to hold the dual reality of a positive intention and a negative impact?

_____

_____

Imagine all of your parts are sitting in a circle around you. You don't need to know exactly who any of them are right now. Can you imagine sending a feeling (or words, if you'd like) of appreciation to all your parts for wanting the best for you? If not, notice which parts might have difficulty with accessing appreciation. Note what this brief exercise was like for you.

_____

_____

_____

Principle 5: Parts interact within our internal systems and with external systems.

In an external system, there are people and institutions that aren't always in harmony with each other. The same goes for our internal systems. Not all of our parts get along. Sometimes they aren't even aware of each other. And sometimes our parts aren't aligned with other people's parts or the larger dominant culture's expectations.

Our internal systems of parts are always getting information from what's happening outside of us. Many of our parts assume their roles in reaction to what happens in the outside world. They may react to other individuals' parts or internal systems, or they may react to larger forces at play, such as social conditioning based on dominant culture norms. To others, *we* are an external system. *Our* parts affect other people. Because we can never know what's going on inside another human being, it's easy to forget that they have parts too. Instead, we may fixate on one part of that person, assuming that simply *is* them. None of us wants to be defined by just one aspect of who we are. Remembering that nothing (and no one) is exactly how they appear to be on the surface can give us all more freedom to be nuanced in our humanness.

Pause and reflect: Bring to mind someone you're fond of. How easy is it to imagine that this person has different parts?

_____

_____

Bring to mind someone you have neutral feelings toward. How easy is it to imagine that this person has different parts?

_____

_____

Bring to mind someone who you find challenging to be around. How easy is it to imagine that this person has different parts?

_____

_____

Think of a part of you that has been greatly impacted by external systems. How did that part come about?

_____

_____

Principle 6: Self has the capacity to heal parts.

There's no magic cure outside of us that can make us whole. According to IFS, what we're seeking is in us already. What our parts need to heal or shift into more harmonious positions is *us*, meaning they need Self energy. Self-to-part relationship is seen as the vehicle for healing. Self energy offers the kind of presence that allows parts to trust and open up. And when that happens, a relationship between Self (us) and parts is possible.

This isn't to imply that we shouldn't reach out for external help or support or that we should be 100 percent self-reliant, and it certainly doesn't mean that we are to blame if we haven't figured things out already. It just means that there's good news—that all of us have an innate capacity to provide care to ourselves, and that care can help us heal painful wounds. We have internal strengths and qualities that can guide us in our healing. I truly believe that we, as queer and trans people, are more resilient and resourceful than we even know.

Pause and reflect: Our society is very focused on finding quick fixes to problems; we're often sold these things with the implication that what we need is outside of us, that we ourselves are somehow deficient. What's it like to consider the possibility that what will allow growth and healing already exists within you and that you just need help or practice accessing it?

_____

_____

_____

What are your reactions to the idea that accessing Self energy from within might help you know when to reach outside of yourself for additional support?

_____

_____

_____

_____

The Goals of IFS: Making Them Your Own

Now that we've reviewed some of the core principles, let's talk about the goals of IFS. I invite you to identify the goals that are most important to you in your growth process. At the end of this section you'll have an opportunity to reflect on how these goals apply to your life.

Goal 1: Unblend and differentiate from parts.

One of the goals of IFS is to help us differentiate from our parts, to know when a part is present and when it's trying to take the lead. Paradoxically, as soon as we notice that we're blended with a part (which is hard to do at times!), we've started to unblend. We then have a chance to be *with* rather than *in* a part of us. The capacity to notice, observe, or zoom out is characteristic of Self energy.

It isn't realistic (or even desirable!) to be unblended every single minute of the day. We can just try our best to notice when we're blended in ways that don't feel right for us or when we feel that we haven't had a choice in the matter. Being blended isn't inherently bad, but when we act in ways that aren't in line with who we are or what our values are, we're probably blended. IFS therapist Natalie Gutiérrez warns us to not weaponize Self; that is, the expectation of being Self-led shouldn't be used to shame people for being blended, especially in dynamics of power, privilege, and marginalization. There are a lot of good reasons why we stay blended, one of which is safety. If we find ourselves demanding that other people be unblended, it's likely that we're blended ourselves! We can aim for unblending while expecting or even welcoming imperfection.

Pause and reflect: How does the goal of unblending feel relevant to my goals/life? What do I want this to look or feel like?

_____

_____

_____

_____

Goal 2: Restore trust in Self leadership.

As we move through our lives, sometimes we do things that we regret or aren't so proud of. Or perhaps other people in our lives have acted in a way to undermine our self-esteem. What can result is a ruptured trust within ourselves. Parts, unhappy with certain dynamics in our lives, can adopt the belief that it's *us* or *Self* that can't be trusted, when in fact they're upset with or lacking trust in another part of us that we've been blended with. There are some parts that tend to try to take the lead, and they can even seem like they are us. But when we remember that Self energy has the qualities that we need to live in line with our values, we can consider that it's parts that don't trust each other.

A metaphor often used in IFS is the idea that we're on a bus with our parts. Ideally, to be Self-led, we ourselves are driving the bus. But sometimes (because of trauma, fear, or other reasons), parts take over and start to drive. Sometimes it goes okay or even really well. We get to where we're trying to go safely and no one gets hurt. However, sometimes a part can commandeer the bus and drive a little recklessly, or maybe a little too slow, or hit the brakes (this might be when someone goes into a kind of emotional shutdown in order to avoid difficult emotions or circumstances). There might be some negative consequences. When things go awry, ideally Self is restored and returns to the driver's seat. Does this mean that any parts get kicked off the bus? No, not at all! We can invite them to sit somewhere on the bus, even somewhere close to us in the driver's seat. And we can take their perspectives into account, but ultimately we're the ones with our hands on the steering wheel. As we get to know our system of parts, we'll become more and more Self-led. And as that happens, we'll learn to trust ourselves better.

Pause and reflect: How does the goal of being Self-led feel relevant to my goals/life? What do I want this to look or feel like?

_____

_____

_____

_____

Goal 3: Liberate parts from extreme roles.

Some of our parts adopt their roles for very good reasons, often believing they have no other choice. Some parts take on extreme roles or behaviors. Often they don't think there's anything extreme happening; they believe what they're doing is absolutely necessary. The trouble is, parts in extreme roles don't tend to allow any space for Self or other parts. Parts in extreme roles work really hard and are often exhausted! We can't force parts to just stop what they're doing, but if we take time to get to know them, we might be able to offer them other options. They can still use their skill sets, ones they've honed over time and used in many circumstances, but perhaps it might feel or look a little different.

If we can become more and more Self-led (and allow these parts to know that we're here for them and can help!), they may be able to access the freedom and choice to take on more balanced roles in the system. Oftentimes, parts with extreme roles are protecting more vulnerable parts of us. If we can build enough connection and trust with the parts that are compelled to work so hard, they might let us get in touch with those vulnerable (and often younger) parts that desperately need our attention. As we help these vulnerable parts heal, their protectors tend to soften and be open to shifting. Parts can feel so much relief when this is possible, and we see tangible improvements in our lives.

Pause and reflect: How does the goal of liberating parts from extreme roles feel relevant to my goals/life? What do I want this to look or feel like?

_____

_____

_____

_____

Goal 4: Increase internal balance and harmony among parts.

As you can imagine, parts don't always see eye to eye, and this can cause a lot of internal tension within us. We end up feeling conflicted or even frozen, not knowing which voice to listen to. This can really create some challenges! As we become more Self-led and help our hardworking protector parts relax more or take on more balanced roles, we can experience a greater sense of internal harmony. All of our parts know that they're part of a collective within us, and they can all rest assured that we're taking care of them. This doesn't happen overnight, but over time, with a practice of working with your parts, you'll experience a difference. A lot of that comes from being able to access the Self energy perspective that every part of us is to be welcomed, heard, and valued. When parts feel like they don't have to fight to get attention (or power), they often feel more secure and the whole system can feel more balanced.

Pause and reflect: How does the goal of internal balance and harmony feel relevant to my goals/life? What do I want this to look or feel like?

_____

_____

_____

_____

_____

Goal 5: Bring Self energy to external systems and toward collective liberation.

Individual healing practices aren't just for us. As we become more internally balanced, this frees up our capacity to care for others and the larger world around us. It makes sense, doesn't it? If we (and our parts) have our emotional needs met, we can shift our focus to what else is going on outside of us and to offer Self energy to people and situations that are in need. It's like the idea of "filling the well"—when we're well resourced, we have more to offer others. When we're less burdened, we can pay better attention to the collective burdens that exist outside of us. Our individual healing and wholeness can then go toward Self-led activism and advocacy for collective liberation. The skills we learn to relate to ourselves can truly have an impact on how we treat other people or how we use our agency in larger systems of inequity. And as we offer care to people and systems around us, we in turn build a greater capacity for Self energy. Just imagine what could be possible if more people were Self-led or could benefit from aligning with Self energy of others! I believe the possibilities are infinite.

Pause and reflect: How does the goal of building the capacity to contribute to collective liberation feel relevant to my goals/life? What do I want this to look or feel like?

_____

_____

_____

_____

_____

In the previous sections you've had the opportunity to consider some of the ways that the goals of IFS may apply to you. In the following worksheet, you can consider what else feels important to you. A copy of this worksheet can be found at http://www.newharbinger.com/55282.

Worksheet: Identifying Your Goals

What are your goals for how you and your life might feel as a result of becoming more Self-led? Be as specific as possible.

1. _____
2. _____
3. _____
4. _____
5. _____

What do you imagine could be possible for yourself if you were able to become more Self-led and in harmonious relationship with your system of parts?

_____
_____

What do you imagine could be possible for queer and trans communities if more people were able to access Self leadership?

_____
_____

What do you imagine could be possible for the world if more people were able to access Self leadership?

_____
_____

To explore your internal system even further, we're going to play with the metaphor of being on a bus with your parts. Set aside about ten minutes for the following meditation, and make sure you're in a relatively comfortable place. You can use this exercise whenever you'd like to help you start to acknowledge parts before having deeper conversations with them. An audio recording of this meditation can be found at http://www.newharbinger.com/55282.

Meditation: The Bus

1. Imagine a bus or vehicle that's large enough to contain you and all of your parts. I'll call it a bus, but it can be a plane, a spaceship, or anything else that you fancy. Imagine that this bus is really "you"—perhaps imagine it in colors you like with whatever features you want it to have. (A hot tub? Movie theater? Why not?!)

2. Imagine sitting in the driver's seat. Try as best you can to ground in this driver's seat while also continuing to notice all the parts that you're currently aware of. Keep in mind that parts show up in all sorts of ways, sometimes in human form, sometimes not. Sometimes they're animals or objects or something without physical form.

3. Take your time to connect with the steering wheel or control center of your bus or vehicle. If at any point you lose connection with this sense of Self leadership, that's okay. Just take whatever time you need to be able to connect to it again.

4. Now notice what's happening with the parts around you. Notice if any parts are trying to take control of the steering wheel or fighting over it. Notice if there are parts sitting quietly, and notice if there are parts that are making a scene! Notice any parts that are judging the entire scenario. As you notice, see if you can stay in a vantage point from the driver's seat.

5. Right now we aren't going to get to know each and every part, but you can start to make some contact and send some Self energy toward them. If you can, imagine sending a feeling of care toward each and every part of you on that bus, regardless of what that part is saying or doing. Let every part know that you're aware of them and that you're interested in getting to know them better. Let them know that you've got the steering wheel and that you're a pretty good

driver. And if they don't believe you, don't try to convince them. Just let them know that over time you'd like to better understand why they feel the way they do. Take a couple of minutes to be in the practice of offering care, recognition, and witnessing to all of the parts that are visible to you today.

6. Before closing, see if you can send appreciation to each and every part for allowing you to be with them today.

7. When you feel ready, you can let go of the imagery brought up in this exercise while maintaining the feeling of being the one leading your internal system and in touch with your parts.

8. Feel free to journal about or draw anything that came up for you in this meditation.

Pause and reflect: What came up for you during the meditation?

_____
_____
_____
_____
_____
_____
_____

Summing It Up

In this chapter you've gotten an opportunity to learn about the principles of IFS and identify your own goals for healing. As you move through the rest of this book, you may want to refer back to your goals periodically, as they may shift. In the next chapter, you'll learn more about connecting with the different qualities of Self energy, a necessary component of healing.

Chapter 3

The You-est You: Tapping into Self Energy

By now you may have a basic understanding of what Self energy is, but you may be wondering how to access it. The exercises in this chapter aim to help you connect to your own version of it. Self energy is universal, but exactly how it looks, feels, and is experienced will vary from person to person. There's no "right" way to have or express Self energy. As queer and trans people, we've been pressured to assimilate in so many ways that simply don't fit for us. So why would we want to assimilate to someone else's version of Self energy? We'll start with some common characteristics of Self energy, and then you'll be able to identify additional qualities that feel true to your experience.

Self Energy: The Eight Cs and More

Self energy refers to who we truly are, the person we know ourselves to be deep down when we're aligned with our values. Self is not a part, but it has the capacity to hold space for all of our parts. Richard Schwartz identified eight qualities of Self energy, also known as the "Eight Cs": Calm, Confidence, Clarity, Curiosity, Courage, Connectedness, Compassion, and Creativity. I offer definitions of each of these as well as two additional Cs of Collectivism/Community and Choice/Consent. Feel free to use a journal or just some paper to reflect on and draw some of the following prompts (no artistic skill required!).

Calm

A sense of spaciousness, peace, or quiet inside ourselves. Groundedness even in the face of stressful situations or parts.

Pause and reflect: Think of a time when you felt calm. What did that feel like in your body? What allowed you to access that feeling?

_____

_____

What situations and conditions help you access a sense of calm? What do you like to do when you feel calm?

_____

_____

_____

Draw whatever comes up as you tap into a sense of calm.

Confidence

Belief in yourself and your strengths. Knowing that you have valuable things to contribute. Trusting that even when mistakes are made, you're valid and worthy.

Pause and reflect: Reflect on a time when you felt confident. What did that feel like in your body? What helped you feel confident?

_____

_____

What do you feel most confident about? How did you gain confidence about it?

_____

_____

_____

Draw whatever comes up as you tap into a sense of confidence.

Clarity

Maintaining a clear, undistorted view of situations and parts, with an absence of projections. The ability to maintain perspective.

Pause and reflect: Reflect on a time when you felt a sense of clarity. What did that feel like in your body?

_____

_____

What do you do to help you gain clarity or perspective?

_____

_____

_____

Draw whatever comes up as you tap into a sense of clarity.

Curiosity

Genuine openness or wonder. A desire to understand without an agenda to change our parts or others.

Pause and reflect: Think about something or someone that you're curious about. Where do you feel this curiosity in your body? Does it feel more from a thinking place (your head) or a feeling place (your heart)?

_____

_____

_____

Draw whatever comes up as you tap into a sense of curiosity.

Courage

Facing challenges or fears intentionally. Resisting injustice. Owning and apologizing for any negative impact of our parts' behavior on others.

Pause and reflect: Think about a time when you did something that required some courage (even if it was scary!). What did that feel like in your body?

_____

_____

_____

Where did you learn to be courageous? Who has modeled this for you?

_____

_____

Draw whatever comes up as you tap into a sense of courage.

Connectedness

Feeling some relatedness or bond within ourselves and with others or a desire to reconnect. Awareness that separateness within ourselves and from others is an illusion.

Pause and reflect: Think about a time when you felt connected to yourself or someone else. What did that feel like in your body?

_____

_____

What activities, people, places, or things help you feel connected to yourself?

_____

_____

_____

Draw whatever comes up as you tap into a sense of connectedness.

Compassion

Feeling care or empathy for the parts of us and others who are suffering. A desire to help without taking on another's pain.

Pause and reflect: Bring up a situation in which you felt compassion for yourself or someone else. What did that feel like in your body?

_____

_____

How do you know that you're feeling compassion without needing to fix, rescue, or take on someone else's pain?

_____

_____

Draw whatever comes up as you tap into a sense of compassion.

Creativity

Feeling connected to creative potential; ability to express oneself in novel or spontaneous ways, without being held back by fear or shame. Openness to inspiration.

Pause and reflect: Think of a time when you felt creative or inspired. What did that feel like in your body?

_____

_____

What situations or conditions help you feel creative? What are things you like to do when you want to express creativity?

_____

_____

_____

Draw whatever comes up as you tap into creativity.

Collectivism/Community

The capacity to be caring toward all living beings and to act in line with this care. Awareness of how our own liberation is inherently connected to the liberation of others, including people who experience different challenges or marginalizations from our own.

Pause and reflect: Reflect on when you've felt a sense of collective care or liberation. What did that feel like in your body?

_____

_____

What conditions help you access a sense that you're part of a greater collective experience?

_____

_____

Draw whatever comes up as you tap into collectivism or community.

Choice/Consent

Awareness that we and our parts have the right to make choices. Honoring boundaries within ourselves as well as those of others. The capacity to recognize the options available to us.

Pause and reflect: Reflect on a time when you felt like you had access to choice or consent. What did that feel like in your body?

_____

_____

How do you know when you're feeling access to your choices and your capacity to give consent?

_____

_____

Draw whatever comes up as you tap into a sense of choice or consent.

Using Creativity to Define Your Self Energy

Part of what I love about queer and trans communities is that we're endlessly creative when it comes to language and how we define ourselves, our attractions, and our relationships. If what's offered by dominant culture (or LGBTQ+ communities) doesn't fit us, we create our own labels and terms that are more affirming. Similarly, there are lots of words (that don't start with a c) that can be experienced as Self energy. Here are some other terms that I've heard associated with Self energy:

Balance

Core Wisdom

Highest Self

Patience

Spiritual Center

Truest Self

Warmth

Wise Leader

The "You-est You" or "Me-est Me" (my personal favorite)

Your experience of Self energy may also be informed by your spiritual or religious background or practices. Some people might refer to Self as the soul, Buddha nature, universal oneness, Higher Power, the divine, or universal love. If you have language that works for you, feel free to incorporate that into your practice of building connection with your parts. However, if you've internalized any religious deities or entities that are oppressive or not affirming of who you are, feel free to leave those out! Many people who have experienced religious abuse or oppression have developed burdened parts or protectors in response, and these are parts that need help. They aren't parts that should be expected to lead the system or your healing process.

Our body's innate wisdom could also be seen as an attribute of Self. It's easy to go through life becoming more and more disconnected from our own bodies and instincts. The concept of Body Trust can be a component of Self energy. Body Trust is the capacity to listen to your body, honor its inherent wisdom, and offer compassion to it and yourself in the face of a world that doesn't treat all bodies equally (Kinavey and Sturtevant, 2022). Body Trust is our birthright, something we're born with that we lose touch with as we learn familial, cultural, or societal rules about what a "good body" is, does, expresses, feels, or looks like. Reclaiming a sense of Body Trust could be considered a form of reclaiming access to Self energy.

We can also draw from our own cultural reference points to inform our Self energy. This may include our hobbies, special interests, pop culture, or other aspects of media (e.g., television shows, movies, music, books, comics). As an '80s kid, I watched the *Care Bears*. In these cartoons, the multicolored bears with different qualities sent a "Care Bear stare" of rainbow light from their bellies to help heal people or objects that had been taken over by a villain's "uncaring magic." Their "weapon" was their care, which allowed people to be transformed back to their natural states of wholeness and goodness. I love this as a metaphor for Self energy and the IFS model of healing through compassion. You're welcome to adopt the Care Bear stare as a reference point for yourself, and I invite you to consider what other concepts you can borrow from your own cultural reference points. The following worksheet will help you reflect on your own experience of Self energy and identify conditions that have allowed you to access it. A copy of this worksheet can be found at http://www.newharbinger.com/55282.

Worksheet: Your Experience of Self Energy

Reflect on which of the "Cs" feel closest to your own experience of Self.

_____

Reflect on whether there are any "Cs" that feel harder to access.

_____

What other words or concepts do you want to include in your version of Self energy? Write down anything that comes to mind, no matter how weird or silly it may sound. Remember, this is for you and no one else!

_____

_____

_____

_____

Reflect on experiences, activities, and times in your day or life where you've experienced Self energy or some of the eight "Cs." Write or draw your reflections.

_____

_____

_____

_____

_____

_____

You can use the next worksheet to track when you're in the eight+ Cs (or use your own words) or feel Self energy in different situations. Keep in mind that intentionally focusing on these qualities of Self energy isn't to push out challenging feelings or experiences in our day-to-day lives, but to make room for the full spectrum of experience, which can include feeling more positive states or experiences. Please keep in mind that there's no achievement in being more Self-led or connected to Self energy. This is only a tool to increase awareness. Try to complete this for at least three different days this week. A copy of this worksheet can be found at http://www.newharbinger.com/55282.

Worksheet: Noticing Self Energy

Day/Date: _____

Quality	Did I experience this today? If so, when/where?	What allowed me to feel this?	What got in the way of me feeling this?	What's one simple thing I could do to access more of this quality?
Calm				
Confidence				
Clarity				
Curiosity				
Courage				
Connectedness				
Compassion				
Creativity				
Collectivism/ community				
Choice/consent				

Summing It Up

In this chapter you've gotten to deepen your sense of Self energy and relate to it more personally. You may refer to this section at any point when you'd like support in feeling more connected to yourself and your capacity to witness your internal system of parts. In the next chapter, you'll learn about protector parts and how to start connecting with them.

Chapter 4

Protector Parts

Protector parts are just what they sound like—parts that have developed strategies for maintaining safety and survival. In times of overwhelm or distress, these parts felt compelled to take on their roles. Some of them feel like they never had a choice but to become responsible for taking care of us or keeping us from feeling overwhelmed. Many of our protectors took on their roles at a very young age, and they themselves may be very young. Sometimes I imagine protectors as little kids dressed in adult clothing.

Beyond taking care of us, protectors shield the parts of us that are more vulnerable, often younger, often the ones that are carrying the weight of having experienced pain or trauma. Protectors operate under the mantra: "Never again." Or they may be protecting us from feeling the pain or overwhelm that our exiled parts are carrying. Our protectors truly help us survive. In this chapter you'll learn more about the different kinds of protector parts: managers and firefighters. You'll also get to try two different methods for connecting with protector parts.

Manager Parts

Managers are parts of us that try to keep things under control. They tend to be proactive and help us cope with our day-to-day lives and stressors. They may do what they do to prevent bad things from happening.

Sometimes managers are praised or socially acceptable, so they feel more inclined to be visible to other people. A manager part that tries to anticipate and say what others want to hear can lead to some desired outcomes (e.g., being liked or seen as agreeable). As queer and trans people, many of us have managers that have helped us fit in, stay connected with others, or prevent social rejection. However, sometimes this comes at the cost of authenticity.

Some managers aren't appreciated by other people. For example, if you have a part that tries to be helpful to others, that can be seen as desirable (e.g., generous, supportive). However, in the extreme, this could look like offering unsolicited advice or telling other people what's best for them. This won't always go over so well with others and can activate parts in them that feel irritated or disempowered.

Different manager parts might:

- try to do things perfectly or always get it "right"
- spend disproportionate time in front of the mirror to make sure we look okay before leaving the house
- try to control a loved one's (or anyone else's!) behavior
- judge us or other people
- go on a diet or restrict food to lose weight
- compulsively say "yes" to other people's requests
- avoid going to an event to avoid social interactions
- act like a know-it-all or try to sound smart to others
- make sarcastic or passive-aggressive comments
- laugh in uncomfortable situations
- make excuses for someone else's unacceptable behavior

Consider this example. Stef's pronouns are they/them, but other people often use the wrong pronoun and refer to Stef with he/him pronouns. Even though Stef feels immediate discomfort when this happens, they have a part that jumps in to keep them quiet. Stef feels a lump in their throat. This part says things to them like, "Don't make a big deal out of it" and "Other people won't get it, so don't even try." This part is concerned that if Stef says anything, they will make other people uncomfortable or be judged as weird. This manager part is well-meaning, but the result is that Stef now feels shame (blending with an exile part) for not speaking up.

Pause and reflect: What are two manager parts that you're aware of in your own system? How are they helping you? How might they be creating challenges for you?

_____

_____

_____

Firefighter Parts

Firefighter parts, like manager parts, develop their roles in order to protect us. However, they're more reactive than proactive. When in a pinch or crisis, they jump in to rescue, distract, numb, or take (sometimes extreme) action. Firefighters often see themselves as the superheroes of the system, responding in any moment when there's perceived imminent danger. Their behaviors can appear impulsive, sometimes even destructive. While firefighters are often deemed extreme, they believe what they do is absolutely justified. They take a "by any means necessary"[iii] approach. Firefighters really aren't concerned about the long term, and at times they couldn't care less about maintaining their image or relationships. They're very much about coming to the rescue in the moment. Sometimes they take over when manager strategies don't seem to be working effectively to protect the system or keep exiles hidden. I imagine firefighters saying, "Move over, I've got this!"

Unfortunately, firefighters often get a bad rap! They're often blamed or disliked by other parts in the system and by other people in our lives. And it's true that firefighter behaviors may create undesirable consequences, even when they have positive intentions. These positive intentions are less apparent when consequences are particularly negative or harmful. I really love firefighter parts because they're lifesaving. I didn't always feel this way, though, as I was blended with my manager parts that just wanted them to behave. If you don't like one of your firefighter parts, check to see if that reaction is coming from another (likely manager) part.

Firefighters are often engaged in the behaviors that we feel we can't control, sometimes things that feel addictive or escapist. Compulsive substance use and self-harm are examples of things firefighters might engage in to quickly douse feelings that are intolerable. Firefighters can

---

[iii] This term is attributed to Malcolm X in reference to achieving freedom and justice by any means necessary. As applied to firefighter parts in the IFS model, I think of firefighters truly believing that what they are doing is for a just cause. Sometimes their behaviors are absolutely warranted and in line with our values!

show up intensely in the body (often like adrenaline or nervous system activation) with "big" energy that we can't ignore.

Some firefighters are socially acceptable. For example, a firefighter who works compulsively to avoid feelings could be praised as "productive" or "hardworking" by a workplace that doesn't value work-life balance. A firefighter who engages in compulsive exercise to compensate for eating what's perceived as "too much"[iv] could be seen as "disciplined" when, in actuality, it's in response to cover deep pain. Some firefighters may even engage in meditation or reading (seemingly harmless activities) to keep us from feeling discomfort.

Sometimes firefighters' behaviors on their own may be neutral, but when firefighters feel compelled to use these strategies for protection, it may lead to feeling out of control or acting in extreme ways. And it may not involve consent from the entire system.

Different firefighter parts might:

- dissociate when overwhelmed
- explode in anger
- zone out with video games
- use substances to numb overwhelming feelings
- have a panic attack in stressful situations
- engage in self-harm
- binge eat or exercise compulsively
- shop compulsively
- shut down suddenly when others are angry
- verbally attack people on the internet
- engage in compulsive sex or porn use
- use work to avoid difficult feelings

[iv] I don't endorse the concepts "overeating" or having "too much" food, as these are not in line with an anti-diet, fat liberation approach. However, I include this because society influences many of us and we (i.e., our parts) may develop these kinds of biases.

Let's return to the example of Stef. After being mis-pronounced, Stef feels an increase in both shame for not speaking up and dysphoria for not being seen as who they are. When they get home and are greeted by their partner, Chi, they try to share about some of the challenges of their day. Chi is preoccupied with cooking and doesn't seem to be listening. Noticing this, one of Stef's firefighters immediately jumps in and picks a fight with Chi, raising their voice and accusing Chi of being selfish. Then another firefighter leads Stef to go to the bedroom, shut the door, and come up with all the reasons why they should leave the relationship.

This is a pattern for Stef—being blended with a part that bolts when there's relationship conflict. Stef knows this about themselves, though, and after about thirty minutes, when they can think more clearly, they realize they were blended. And this realization allows them to further unblend and access Self energy. They access courage and a desire to be connected, and they go back to the kitchen where Chi is and apologize. Stef realizes that this is a pattern for them in relationships, so they really want to do some work to heal the part of them that gets so scared that another part (firefighter) urges them to flee. Chi, experiencing the Self energy of Stef, is able to access Self energy as well and apologizes for not being more present. Stef and Chi are able to give each other a hug and feel more connected again.

Pause and reflect: What are two firefighter parts that you're aware of in your own system? How are they helping you? How might they be creating challenges for you?

_____

_____

Balanced versus Burdened Protection

We all have protectors that help us or bring strengths to different life tasks and situations. However, these parts' strategies aren't always helpful or in line with who we know ourselves to be. So, what's the difference between a protector who's generally helpful versus one who's causing us or others some problems? The key difference is that the latter is typically burdened, meaning that it carries a sense of responsibility and a lack of choice about its role in the internal system. A burdened part doesn't believe things will be okay unless it does whatever its job is. It doesn't believe Self energy can do the job to protect the system.

A key question to help us better understand a protector and what/who they're protecting inside us is: **"What are you worried would happen if you stopped doing what you're doing?"** This question is really important because it gets to the heart of what the part is protecting. Whatever fear or concern it shares points to whatever parts (and related feelings, sensations, or beliefs) have been cast out of conscious awareness. In other words, a fear that the protector names is key in helping us uncover the exile that it's protecting. When it comes to queer or trans experience, a very common dynamic is having protectors who want to maintain physical safety or connection with others, often at the expense of one's truth. Sometimes this is related to past trauma, but sometimes it's a very real present-day conflict because of ongoing oppression. It's important to take time to get to know and understand protectors' fears and ask for their consent to connect with exiles. We'll learn more about exiles and how to work with them in chapter 6.

A Note about Polarizations

The concept of a polarization was mentioned briefly in chapter 1. Polarizations occur when two or more parts seem to be in tension or conflict with each other. The most common polarization is one between a manager and a firefighter, but this isn't true across the board. In a polarization, it's common to be blended with both/all parts or to vacillate between being blended with one side of the polarization and then with the other side(s) of the polarization. If you identify a polarization, you'll want to work with one part or side of the polarization at a time. Let your parts know that you're interested in hearing from both/all sides, and then use the following exercises of getting to know protectors as you hear from them just one at a time. You may find that, if you're open to hearing from both/all sides, that your parts' concerns or goals may not be so misaligned. Remember, managers and firefighters may have very different strategies even when their motivations may be quite similar (e.g., to protect a vulnerable exile).

Connecting Directly: Centering a Part

When connecting with parts, it's best to speak with one part at a time and to know who you're speaking with. Traditionally in IFS, we use the term "target part" to refer to the part that we've chosen to start with or focus on in any given moment. I've chosen to use the term "**centered part**" instead because those of us who belong to (sometimes multiple) marginalized groups have been targeted in a negative way. If the word "target" works for you, you're welcome to use it. My intention here is to offer an alternative. The following sections offer two different methods for having a conversation with a protector part. If one or both of these methods feel helpful, you can do them on a regular basis whenever you want to get to know a part better.

Method 1: The Six Fs Conversation with a Protector

This method uses an approach called the "Six Fs" to connect with a protector part. The six Fs are: Find, Focus, Feel Toward, beFriend, Flesh Out, and Fears. In a nutshell, this process is about making sure that there's an adequate Self-to-part connection and getting to know a protector's role, possibly how it came to be, and its fears or concerns.

A few tips: 1) Take your time and be patient, as some parts communicate more slowly than you might expect. 2) If after trying this for a while the practice feels impossible, don't give up. Take a break and come back to try again in a couple of hours or days. 3) If you give this method a second try and it still feels impossible, skip this section and go to the next one. That method approaches a conversation with parts from a different angle.

You'll want to set aside at least fifteen to thirty minutes for this exercise, if not more. Choose one **protector** part that you'd like to center or get to know better. It can be a manager or firefighter, and if you're not sure which it is, that's okay. You can still get to know it. A copy of this worksheet can be found at http://www.newharbinger.com/55282.

Worksheet: A Self-Led Conversation with Protectors

Find: Which part would you like to center first?

- How would you like to refer to this part? If you're going to come up with a name/label for this part, try to do so from a Self-led place, ideally without judgment. It can be simple and descriptive, such as "the one that blows up" or "the busy part."

  _____

- How are you experiencing this part in or around your body?

  _____

Focus: As you focus on the centered part, what else do you notice?

- What sensations do you notice, and where do you experience them?

  _____

- What feelings do you notice?

  _____

- What thoughts or beliefs come up?

  _____

- Are there any images that come to mind?

  _____

- Is there any other way that you're experiencing this part (e.g., texture, color, temperature)?

  _____

- Is there anything else you notice about the part in terms of what it's saying or doing or how it's relating to you?

  _____

  _____

  _____

Feel Toward (check for Self energy): Check to see if there's a critical mass of Self energy available. How are you feeling toward the part? How open do you feel to spending time with it?

_____

_____

beFriend: Are you able to approach the part with Self energy?

- If you're able to access any care, compassion, or curiosity, let the part know that or send it one of those qualities of Self energy. You might energetically send a feeling of warmth, care, or compassion. Or you can say to the part, "I'm curious to know more about you." Or you can simply say to the part, "I see you. I'm here." How does the part respond to your communication?

  _____

- If you notice any other reaction that doesn't seem in line with Self energy (e.g., judgment, fear, impatience, or rushing), that's okay. It means that there's another protector part present. Ask this second protector if it would be willing to give you some space so that you can be in contact with the centered part. Let it know that it doesn't need to change how it's feeling, and it doesn't need to go away. Just see if it would be willing to give you a little space, even a few minutes, without interrupting.

- If the part that jumped in doesn't want to unblend, ask it: "What are you concerned about? What are you worried might happen if I give attention to the centered part?" Hear its concerns and try your best to validate that it's feeling the way it does for a reason. Then see if it might give you some space. Is the part now willing to unblend?

- If the part is still unwilling to unblend or let you spend time with the initial centered part, then this part (the one that's jumping in, judging, etc.) becomes the centered part. Start this exercise over with this second part as the centered part (the one you're focusing on). Write down what your new centered part is. You may have to do this a couple or several times until you find a part that you can easily be in a Self-led conversation with.

_____

_____

_____

_____

Flesh out: When you feel enough Self energy present, ask the centered part one of the questions below, pause to listen for an answer, and notice anything else about the part and its connection to you. In the lines below, jot down whatever comes up.

- What's your job?
- What do you like about your job?
- What do you dislike about your job?
- How do other parts respond to you?
- How do other people in my life respond to you?
- How did you come to be in this role?
- If you didn't have to do this job, what would you choose to do instead?

- _____

- _____

- _____

- _____

- _____

Fears (key question): Ask, "What are you worried would happen to me if you didn't do this job?" Make sure to note any concerns that the part shares.

_____
_____
_____
_____
_____

Continue to befriend: As the part shares with you, see if you can offer understanding. Don't force it if that's not what you're feeling.

- If it does feel genuine, you can say things like, "It makes sense that you'd feel the way you do," "I hear you," "I really get what you're saying," and "Is there anything else you'd really like me to know?"

- When it seems that the part has voiced all it would like to share (for now), thank it for sharing with you. Then check to see how connected you feel to the part and how the part is responding to you. Has anything shifted between the moment you noticed the part and now? Jot down any new insights or shifts you're aware of.

_____
_____
_____
_____
_____

Method 2: Direct Dialogue with a Protector

This is another way to invite a conversation with a part. Some people (especially those of us who are overthinkers, i.e., have thinking parts who often take the lead) respond to this method better because it involves free flow or stream-of-consciousness writing. In this method, you'll be shifting perspectives, meaning you'll write from the perspective of yourself as well as allow a centered part to write for itself. If you have parts that are skeptical about this, ask those parts to suspend disbelief for this exercise and allow you to even *pretend* that you're the part. There's no right or wrong in this method; just write whatever comes up without thinking about it too much. I find that giving the part full permission to express itself can allow some interesting (and sometimes amusing or surprising) insights to come up. Sometimes having the dialogue contained on something concrete and visual (e.g., paper) also allows my neurodivergent brain to stay more present. You'll need at least fifteen minutes for the following exercise. A copy of this worksheet can be found at http://www.newharbinger.com/55282.

Worksheet: Hearing Directly from a Protector

Get out at least two writing implements with different colors and/or textures. You may want to have colored pens or even more options so that you can go with what feels right in the moment. If at any point you need more space, you can use another piece of paper or a journal. This exercise involves using a timer so that this becomes more of a conversation rather than alternating monologues.

Begin by taking a few deep breaths. Bring to mind a protector part that you'd like to get to know better, one that you feel curious about. Which part would you like to be in correspondence with? Which part would you like to center right now?

_____

First you'll be writing from your own perspective, with the intention of writing from a Self-led place. Set a timer for three minutes. In the space below, write to the centered part or the part you've chosen to communicate with. Let it know that you're interested in getting to know it. Let it know how you experience it, and ask it any questions that you have about it.

_____
_____
_____
_____
_____
_____
_____
_____
_____

When the time is up, switch perspectives, meaning you'll invite the part to be fully embodied and to write for itself. In the moment, give the part choices. Allow the part to choose the writing implement or color. The part may or may not want to write with the same hand as you typically do (e.g., even if you're left-handed, you may have a part that wants to write with your right hand). The part may be less verbal and may want to draw instead of write. The part may also want to move to a different chair or area of the room you're in. Now set the timer for another three minutes and give full permission for that part to express whatever it needs to in the space below.

_____

_____

_____

_____

_____

_____

When the three minutes are up, switch perspectives again, returning to being you and in your own Self energy. Now you'll have an opportunity to respond to the part. You can thank it for sharing what it did. You can ask more questions. You can share your reactions. Just write or share whatever comes up for you. Set the timer again for three minutes and begin.

_____

_____

_____

_____

_____

_____

_____

Protector Parts 65

When the three minutes are up, take a few deep breaths. The conversation has had three passes now (you, the part, and you again). You're welcome to keep the conversation going on another page or in your journal and take as much time as you'd like to learn more about the part through this back-and-forth conversation. When you and/or the part feels complete enough or you sense you're at a stopping point, answer the following reflection questions:

- What was it like to do this exercise? What did you learn?

  _____

  _____

  _____

  _____

- What, if any, differences did you experience in your body as you shifted perspectives?

  _____

  _____

  _____

- Is there any difference in how you feel *toward* the part after allowing it to express itself through writing or drawing? Upon reflection, were you blended with any other parts initially?

  _____

  _____

  _____

Summing It Up

In this chapter you've learned more about protectors (managers and firefighters). You've identified and started to connect with some protectors within your own system. It's crucial that at this stage you really take your time to practice getting to know protectors and not push past their comfort zones. Remember, IFS is deeply embedded in consent and respecting the protective system. In fact, you may even take a week or two to practice getting to know protectors every day before you advance to the next chapter, which will focus on the process of unblending with your protector parts as you connect with your own system or interact with other people.

Chapter 5

Unblending While Respecting the Protective System

In the last chapter you were guided through conversations with protector parts. In some ways, being able to communicate at all with our parts requires some amount of unblending. This chapter provides some additional support with unblending so that you can access more Self energy.

Blended versus Self-Led: A Continuum

Being "blended" isn't all or nothing; we can be totally blended (parts-led), not blended (Self-led), or somewhere in between (some combination of Self energy and parts energy). Approaching the concept of being blended versus Self-led as a continuum can allow us to work with what we've got, as well as work with what's realistic.

Say you were having a birthday party and a close friend who said they were coming just didn't show up or communicate at all. You might have a range of responses:

- Being completely blended with a part that feels hurt and brings in memories of all the other times that people haven't shown up for you

- Being completely blended with a part that sends a nasty text to your friend (now ex-friend!)

- Being completely blended with a part that's angry and critical, making a mental list of all of that friend's worst qualities

- Being completely blended with a part that doesn't want to say anything for fear of creating conflict

- Being partially blended with the angry and critical part but also able to access awareness of some of your friend's positive qualities and times when they've shown up for you, but still feeling unable to decide whether and how to respond to your feelings

- Being unblended and able to slow down, identify the different parts who are having reactions, offering witnessing and care to these parts, and then being able to access more Self leadership; this could involve curiosity (is your friend okay?), courage (approaching your friend and speaking for, not from, your parts), or clarity (spending some time reflecting on whether this is a pattern, how important the friendship is to you, and what you might be able to ask for in the future)

Of course, as this is only an example, and your Self-led response may be very different from someone else's based on your specific context. To reiterate: being 100 percent Self-led isn't the goal; having a critical mass of Self energy (i.e., being *mostly* Self-led or having *enough* Self energy to offer to our parts) is really what we're going for.

Permission and Consent

In doing parts work, it's important to be aware of the tension between permission/consent and having an agenda. It's important to check to see *who* inside is trying to facilitate healing. Is it led by Self energy? Is there consent given from other parts in the system? Or is there an agenda or rushing quality, or even a desire to push away any parts that seem to interfere or slow down the process?

It makes sense that parts of us want to help us feel better or at least escape pain. However, when it comes to trauma recovery, there's only so much that our systems can tolerate before getting overwhelmed and either freezing or shutting down. Rushing the process can inhibit building trust. The concept of **titration** can be useful in reminding us that the most effective way to engage in healing work is by doing it in tolerable amounts. If we try to move too quickly, there may also be **backlash** within the system. When one part takes over without respecting other parts' concerns, those parts are likely to resist, rebel, and block the process. With healing work, faster isn't better.

If we're aware of being blended with a part, we can *ask* the part to unblend or give us space. We don't do the unblending; the part is the one that can unblend. Here are some ways that I've respectfully asked parts to unblend:

"Would you be willing to pull to the side and allow me to connect with [part]?"

"Would you be willing to soften back and give some space for me to have a conversation with [part] without interruption?"

"You can stay right here and jump in again if you need to, but for now would you be willing to allow me to talk to [part]?"

A Personal Example

While writing this book, I suddenly developed some serious sciatic nerve pain. Part of me panicked and wanted to urgently stop the pain. That part wanted to go straight to the source and use a massage gun directly where I was experiencing pain on my hamstrings. Much to my dismay, this didn't help!

Shortly after, I learned that applying pressure directly to the site of the pain would actually make it worse. Whoops! What I needed to do instead was place the massage gun on the muscles adjacent to the pain to encourage them to relax. If those muscles could relax, that could help the places where the sciatic nerve was inflamed. I laughed at myself because of course that kind of approach is so socialized in me: *Go to where the problem is and fix it! Use force to try to change something! And make it fast!*

This analogy fits well for protectors and exiles. Part of us may want to go directly to exiles to make the pain stop. In doing so, we bypass doing more intentional and paced work. It may seem counterintuitive to first connect with protector parts, but that's what's needed. Some of my early training as a therapist taught me to address the "first and worst" traumas in a client's life. Years later, I now know that moving too quickly is neither trauma-informed nor body-informed. IFS has led me to respect that the internal system needs time to build trust not only with me as a therapist but also with the client's Self. Over time, protectors become willing to consent to the client's Self connecting with the exiled parts.

It's helpful to spot when a part has an agenda and to ask that part to unblend so things can happen in their own time. Also, the parts searching for a "quick fix" or "productive" way to engage in growth work likely have internalized colonialist and capitalist ideals. These protective parts will need attention from you if they keep trying to manage the pace of your process. If they're willing to collaborate with you, this will be supportive of your entire system.

Pause and reflect: Do you relate in any way to the previous example? Can you think of situations in which parts with urgency or agendas took over in your healing work? What was the outcome?

_____

_____

_____

_____

_____

Addressing Protector Fears

Protectors are often very committed to their jobs. While many of them may feel pride in being good at what they do, they may not always enjoy being in these roles. Oftentimes they're scared that something bad will happen if they stop doing their jobs. Because of a variety of fears, they often interrupt the process of connecting with other parts that need attention. When this happens, we can respond from a Self-led place and gently address their fears or concerns. Here are some common protector fears and how we can offer them perspectives that will allow them to better trust in the healing process. Above all, we want to convey to protectors that what they're feeling makes sense given our life experiences and that they don't have to deal with anything alone. Use whatever language feels most natural and authentic for you; these are just some options.

Protector Fear	Self-Led Responses
Feeling overwhelmed by pain	"I can check in with you through the process to make sure it feels like a tolerable amount for you and the entire system." "You can jump back in at any time if it feels too intense." "Would you be willing to let me work with even 10 percent of what the other part is experiencing?"
Nothing will help, so what's the point?	"I know you've tried really hard to make things better. Would you be willing to try something different and allow me to help?"
Another part will take over and cause damage or harm	"If we're able to spend time with that part, we can help it heal and find a different way to respond to what's happening."
Being judged by others	"We have full choice about what's shared with other people." "No matter what happens, I'm here with you."
Losing its job or being kicked out of the internal system/family	"You'll always have a place here, even if your role changes. You get to decide whether and how anything shifts."
Uncovering painful memories or trauma	"If something new or painful shows up, we can find a way to help whatever parts are carrying that pain so that they don't need to carry it forever."
If healing occurs, the outside world won't support it or be able to handle it	"We don't have control over the outside world, but the healing will allow us to access more choices and be less distressed when things are hard."

You can use the following meditation any time a protector expresses fear, concern, or a strong reaction to another part. Keeping in mind that change can only happen through the foundation of a relationship, take time to really understand and respond to the part's concerns. Be open to having several conversations with protectors who have strong fears or concerns. Set aside at least fifteen minutes for this meditation. An audio recording of this meditation can be found at http://www.newharbinger.com/55282.

Meditation: Respecting Protector Fears

1. Start by taking a deep breath and settling yourself in a comfortable position.

2. Bring up a part in your system that you're aware of but have not spent much time connecting with yet. As you focus more on this part, notice however it's making itself known to you: through images, words, feelings, body sensations, or anything else. Really focus on this part, letting it know that you're aware of it.

3. Now allow that focus to soften a little. You may imagine that you're zooming out, still seeing the part but also holding a wider perspective.

4. Notice if there are any other parts that are having reactions, such as concern, judgment, or even dislike toward the first part. Invite one of these parts to come forward.

5. Take your time now to notice how the part that's having a reaction is showing up or expressing itself to you. Notice images, words, feelings, or sensations that are connected to it. Let it know you'd like to hear its concerns. Invite this part to share what it's worried will happen if you connect with the first part. Try to listen and take in whatever this part has to say. Consider this part's perspective.

6. Thank this part for sharing with you and helping you better understand its perspective. Send it some compassion, letting it know that it gets to feel whatever it's feeling and that it makes sense. See if you can address whatever concerns or fears it expressed. What do you want to say to this part?

7. Notice how this part is responding to your attention and presence. Has anything shifted in the way it's showing up?

8. Letting this part know that you're taking its concerns to heart, ask if it would be willing to give you enough space to have a conversation with the initial part you

noticed without interruption. Let it know that if it gets really scared at any point, it can jump back in, but for now ask it to give you some space.

9. Now shift your attention back to the first part you were interested in getting to know. Let it know you're back, that you'd still like to connect, and that you had to get permission from another part to do that.

10. Invite this part to share anything it wants you to know about itself. Take as much time as you need to hear from it, seeing if there's anything in particular it would like from you. Thank the part for allowing you to witness it.

11. Finally, thank both parts for allowing you to connect with them, making any arrangements to continue these conversations at a later time if needed. When you're ready, take a deep breath and bring your focus outward again.

12. Spend a few minutes jotting down some notes.

Pause and reflect: What came up for you during the meditation?

_____
_____
_____
_____
_____
_____
_____

The following worksheet will help you synthesize what you learned about your protector fears in the previous meditation. A copy of this worksheet can be found at http://www.newharbinger.com/55282.

Worksheet: Exploring Protector Fears

When connecting to the part that was having a reaction toward the first part, what was the main fear or concern that it expressed?

_____
_____
_____
_____
_____

Were you able to respond to the part's concerns from a Self-led place? If so, what did you say or do? Did you notice any shifts in that part as you connected with it?

_____
_____
_____
_____

Is there anything else significant you noticed in this practice?

_____
_____
_____
_____
_____

Speaking for Parts versus from Parts

When we're blended with parts, we aren't just experiencing that within ourselves. We have impacts on the people around us. It's probably not hard to think of a time that you were so blended with a part that you said something you didn't mean or in a way that hurt someone else's feelings. Sometimes our parts are so caught up in their agendas or roles that they don't consider how they affect others, while at other times our parts can be so caught up in how they appear to others that they sacrifice our own needs.

When we're blended, we tend to speak *from* our parts. It's as if those parts have the mic and are speaking for us and the rest of the internal system. When we speak *for* our parts from a Self-led place, it's as if we have the mic and we can make choices about what we share with others and what we don't. This isn't to silence parts, but to allow us to have discernment. Our parts are always welcome to express what they want to express to *us*, but we may choose to have some boundaries when it comes to what's shared with other people. And if we, from a Self-led place, are open to hearing from our parts, then their concerns will be considered and can inform our decisions and actions. When speaking *for* parts, there's a greater likelihood that others will be receptive and that we will be heard. It also allows for us to honor the multiplicity of our experience, as we can speak for a number of parts who have differing valuable perspectives to offer.

There are times when a part won't feel safe enough to unblend. When that's the case, we must respect the part and see if we can better understand its concerns. For many of us who belong to marginalized groups, it feels too vulnerable to enter dominant culture spaces without an entourage of protectors. We don't want to shame ourselves or others for being blended when it's for the purpose of survival or system regulation. We can aspire to be as Self-led as possible, and if that's just not going to happen, we can accept our parts for where they are in any given moment.

Here are some examples of what it might sound like to speak *from* a part versus *for* a part.

Speaking from a part (blended or parts-led):	Speaking for a part (Self-led):
"You don't care about me!"	"I have a part that's holding a story about you not caring about me."
"If I'm real and act like myself, everyone will judge me."	"A part of me is scared that if I'm myself, others will judge me."
"I'm really pissed off at you. How could you do that?"	"A part of me is really angry in response to what you did, and a part of me wants to understand why."

Pause and reflect: Think of a time when you said something when you were blended and it didn't go so well. How might you say that differently from a Self-led place?

_____
_____
_____
_____
_____
_____

Think of a time when you were able to speak from a more Self-led place, even if it was something hard to say to another person. How did that land?

_____
_____
_____

Mapping Parts

Sometimes I joke about having a "popcorn" system of parts. When I first started my own IFS journey, it felt like my parts were popping up left and right like popcorn kernels, and I didn't know which to pay attention to. And as soon as I tried to talk to one, another one would pop up! And then another, and so on. It's gotten easier now that I've stopped expecting myself to have a single focus or to be able to track every single part that shows up. It's still like popcorn in there sometimes, but things have slowed down enough for me to be able to notice more.

Much like in any mindfulness practice, distractions are to be expected. Often these "distractions" are actually the parts that we need to attend to the most! With this practice, you'll become a better "parts detector" with a greater capacity to be Self-led and less blended (parts-led). Remember that everything is connected—to you. If you don't "catch" something the first time it shows up, trust that it will come back around when the time is right. Tracking can help you know which parts to come back to; it can also help with unblending. Here are some methods you might want to try to see what works best for you and your system:

- Make a simple list. Jot down whatever you observe or learn from a part so that you can recognize it when it shows up again.

- Use a chart. Divide a page into seven columns and write down 1) what you first noticed about the part, 2) body sensations, 3) emotions, 4) images, 5) thoughts or beliefs, 6) action urges, and 7) how you feel toward the part.

- Map relationships among parts: every time you notice a part, write down what the part is and draw a circle around it. Then start to draw lines between parts that have a relationship with each other.

- Map the body: draw an image of a body. Then write down parts wherever you're experiencing them in your body.

- Invite your parts to draw themselves or write freely from their own perspective.

- Use external resources. There are more and more resources out there to help with tracking parts, including apps and books. I recommend Michelle Glass's book on daily parts meditation (Glass, 2016).

The following worksheet is just one example of how you might keep track of the parts you meet. There's an example of a completed chart followed by a blank one you can use. A copy of this worksheet can be found at http://www.newharbinger.com/55282.

Worksheet: Chart Your Parts (Example)

Write down what you notice about three different parts that you're aware of right now.

	Part: Stressed about work	Part: Wants to quit	Part: Be logical!
What's the first thing I notice?	I feel stressed	Sense of urgency	It doesn't like the one that's urgent and wants to quit my job
What, if any, **sensations** do I notice in my body?	Tension in shoulders	Heart racing, energy in my chest	Feels like it's in my head
What, if any, **emotions** do I feel?	Overwhelmed		Annoyed and concerned
What, if any, **images** are coming up?	All the work I have to do; stacks of papers	The part looks like a version of me trying to run away	It looks like a person—not me but a man in a business suit. Weird!
What **thoughts or beliefs**, if any, is the part expressing?	"It's too much."	"Let's get out of here!" It's telling me all the reasons why this job sucks.	"You need to pay your bills. You have to think this through!"
What, if any, **urges to do something** am I experiencing?		Urge to quit my job	
How am I feeling **toward** this part? How open do I feel to getting to know it better?	Pretty open	A little scared of doing something impulsive (this is another part)	Curious about all of them
Anything else I notice?			In just writing these things down, I feel the parts relaxing and there's more spaciousness.

Worksheet: Chart Your Parts

Write down what you notice about three different parts that you're aware of right now.

	Part: Stressed about work	Part: Wants to quit	Part: Be logical!
What's the first thing I notice?			
What, if any, **sensations** do I notice in my body?			
What, if any, **emotions** do I feel?			
What, if any, **images** are coming up?			
What **thoughts or beliefs**, if any, is the part expressing?			
What, if any, **urges to do something** am I experiencing?			
How am I feeling **toward** this part? How open do I feel to getting to know it better?			
Anything else I notice?			

Summing It Up

In this chapter you've learned more about unblending from your protector parts, the importance of respecting the pace of the protective system, and common protector fears. You've learned the difference between speaking *for* parts (unblended, Self-led) versus speaking *from* parts (blended, parts-led). We can hold the goal of being Self-led (imperfectly!), but when this isn't possible, we can offer compassion and patience to ourselves and our parts. We've also gone over a few ways to keep track of your parts as you work with your system. The next chapter will examine the other important category of parts: exiles.

Chapter 6

Exiled Parts

Now that you have a better understanding of the protectors, let's talk more about exiled parts or exiles. The term "exiled" doesn't mean that parts ever get banished from the entire internal system, as we can't get rid of any of our parts. Rather, these are parts that are hidden away from our conscious awareness because they're carrying burdens that may be too much for our system to tolerate in a given moment. Burdens are like wounds: they can be the experience of pain, shame, or other intolerable beliefs, feelings, sensations, or qualities. Exiles and their burdens are often unseen, unheard, and unexperienced.

Our protectors may work so hard that our exiled parts aren't fully known to us. We may have some sense that they exist, but sometimes even pausing to think about them will activate a protector to jump in and distract us from getting too close. So most of us, unless engaging in some kind of intentional healing process, won't get a true sense of these parts as we move through our lives. When the strategies of our protector parts fail or are no longer as effective as they once were, we become more aware of our exiles and their burdens. This is often the time that people feel motivated to seek support such as therapy.

How Parts Take on Burdens

Parts may become burdened in response to a wide range of external situations, including being explicitly mistreated, abused, bullied, or neglected. Similar burdens may result when parents are well meaning but their behavior or subtle communications have a negative impact. For example, a parent could be unaware that they offer much more praise when a child conforms to gender norms, which could inadvertently communicate that it's not okay to express anything contrary to these norms. Exiles may internalize the negative belief of "It's not okay to be me." Protectors may then make sure that the child continues to conform with gender norms, repressing any desire to deviate from them. Regardless of the intention, the impact is felt and internalized.

Parts can get exiled when they receive the message that it's not okay to feel a certain way. It could be an explicit message: "Stop worrying so much." Or it could be an implicit message: "Stay positive! Good vibes only!" An exile may take on the belief of "It's not okay to be scared" or even "It's not okay to express my feelings." Sometimes exiles get so accustomed to their pain that they believe they *are* their pain. Protectors hide exiles away because they don't want them getting hurt in the same way again.

The thing is, our exiles deeply want us to witness them. They often want redemption or repair. When protectors are disarmed, exiles use this as a chance to desperately get our attention or flood our nervous systems. If we're unable to access Self energy, a protector will likely step in to keep exiles out of awareness and nothing will change. However, if we can befriend our protectors from a Self-led place and get consent from them to meet our exiles, so much more is possible.

Pause and reflect: Are you aware of any of the negative messages or burdens your exiles have been carrying? What do you notice about the energy of these parts?

_____

_____

Legacy Burdens

Sometimes burdens are passed down through generations, such as from a grandparent to a child. For example, in some families it's acceptable to express certain feelings but not others. Everyone in the family might hold the implicit message of "Anger isn't allowed." Unspoken rules such as these that get passed down in a familial or cultural lineage are what we refer to as **legacy burdens**. A legacy burden is a belief, thought, feeling, behavior, or sensation that was passed down through prior generations and is typically not the result of a person's own lived experience. Sometimes the original event that caused the burden is unknown because it goes back so many generations, especially when there's a culture of silence about uncomfortable topics (which may be the result of a burden itself).

Burdens get passed down by the direct witnessing of another person's behavior, as would be the case in a child who learns how to restrict food from a parent who's a chronic dieter. Or it may be more hidden, such as a child who develops anorexia and is unaware that someone three generations back in their lineage also had anorexia. Sometimes the burden manifests as the same behavior despite having different reasons. In this example, the person's anorexia may be tied to a burden that originated in a time of famine or forced displacement. There's a great deal of evidence for the intergenerational transmission of trauma *at the cellular level* (Lehrner and Yehuda, 2018).

Legacy burdens aren't just passed down through direct blood lineage. A more specific type of legacy burden is a **cultural burden**. This is typically a burden that's held by an entire group of people, often rooted in marginalization, past or present. Diet culture, which is rooted in anti-Blackness (Strings, 2019), leads entire societies to carry the burden of feeling that their bodies aren't good enough. This is just one of many examples of a kind of cultural trauma that can be pervasive and intergenerational.

Many of our wounds and traumas as queer and trans people aren't isolated to our individual lived experiences. In our psyches, many of us carry the burden of knowing that anti-trans and anti-queer structural and interpersonal violence impacts our communities. For those of us who face marginalizations based on race, social class, disability, or some other marker of "difference" from the dominant culture, we may carry multiple burdens based on our identities. Some of us may carry heavier burdens than others. For example, queer people of color, especially trans women of color, experience the highest rates of violence as well as health disparities. People with these identities (e.g., being a person of color, being a trans woman or trans feminine) may understandably internalize these burdens at a deeper level.

The following is a list of other examples of causes of legacy burdens that may result from cultural or ancestral trauma passed down over generations. These subjects can bring up a lot of feelings and often a lot of grief or pain. I offer these examples to highlight how things that we might believe are unique or personal may be rooted in something beyond our power to control. Some of these overlap with each other.

- Patriarchy, sexism
- Deprivation related to famine, war, forced displacement
- Xenophobia, being seen as a perpetual outsider

- Colonialism and anti-Indigeneity
- Slavery and anti-Black racism
- The prison industrial complex, carceral punishment
- Eugenics and intersecting ableism
- Anti-fatness and diet culture
- Neurotypical expectations
- Religious oppressions and persecutions, such as anti-Semitism and Islamophobia
- Sanctioned punishment or persecution in places where being queer or trans is illegal/criminalized
- The AIDS/HIV epidemic, homophobia, and the failure of our government or health care systems to respond accordingly
- Laws and bills that outlaw education about queer and trans people, participation in sports, drag, access to safe restrooms, or gender-affirming health care

You might be thinking, "How can individual healing do anything if these external systems are still causing so much harm?" Even when we can't shift enormous oppressive systems all on our own, or when the harm is still happening because it's pervasive in our society, we can heal our relationship with ourselves. That's not to say we shouldn't let these things get to us; I can't imagine how that would be possible. But we can learn how to better care for our hurt parts so they don't feel so alone or powerless. And we can do this not just in isolated silos, but by connecting with others who might understand what we're going through.

In the next few chapters we'll explore how we can work with parts who are carrying the wounds or burdens of our life experiences or those that have resulted from legacy or cultural traumas. To begin with, I'd like to offer this brief meditation on just making the smallest connection with an exiled part. In the next chapter we'll build the relationship and the capacity to heal these parts' burdens, but for now it's enough to make contact and build mutual Self-to-part awareness. Think of this as the beginning of a longer conversation. I recommend practicing this at least three times before moving on to the next chapter. An audio recording of this meditation can be found at http://www.newharbinger.com/55282.

Meditation: Making the Smallest Connection

1. Begin by choosing a protector you know well, perhaps one you've identified while working through the last couple of chapters. If you'd like, you can revisit the exercises in chapters 4 and 5 to find appreciation for what this protector part does, how it came to be in its role, and what it's worried will happen if it doesn't keep doing what it's doing. Or you can just sit for a few moments trying to make contact with the protector and building a sense of connection with it.

2. When it feels right, you can ask that part if it would be willing to let you connect with whatever part it's protecting. If it's not willing, just stay with it and see if you can hear more about its concerns. If the protector part is willing to step aside and allow you to see the part it's protecting, thank it.

3. See if you can notice the exile part, however it's showing up. Do you notice an image, sensation, feeling, or thought connected to this part? In your mind's eye, notice how close to or far away you are from this part. If you're far away, imagine getting just a little bit closer. Try to embody the gentle energy that you might if you were approaching a small animal who's shy or timid.

4. Is this part aware of you? If not, see if you can get a tad bit, even an inch, closer to it. Do this until the part is aware of you. Let it know that you're here and that you see it. You may speak the simple phrase, "I'm here with you." How is the part responding?

5. Imagine being with the part, without needing to have any words. You could imagine sitting somewhere together, like a park bench or under a tree or anywhere else that feels right. If the part seems pressured or urgent in connecting with you or has a lot to say, let the part know, "I'm not going anywhere. I want to take time to get to know you." Just take in each other's presence. This smallest connection will be the foundation of your relationship.

6. If you can, send this part some care or compassion or however you're experiencing Self energy, and notice how the part responds.

7. Take a few more moments to just sit with the part.

8. When you're ready, you can thank the part for allowing you to even make the smallest connection with you and let the part know that you'd like to return for a longer conversation.

9. Take a few moments to reflect or jot down anything that came up for you in this practice.

Summing It Up

In this chapter you've learned more about exiles and burdens. You've begun to make contact and build some awareness of what's happening internally for these parts of you. In the next chapter, you'll start to learn how to engage in a process of helping these parts heal.

Chapter 7

How Parts Heal

Healing is neither a simple concept nor a simple process. Definitions of healing vary widely and are often influenced by the biases of the world. Ableism tells us that people or bodies without pain or trauma are more valuable and that we all have equal access to attaining health. If we approach healing in a perfectionistic way, we may think it means having no "issues" left to work through. We become susceptible to self-blame for not being healthy or healed, leading managers to believe that they aren't working hard enough to "achieve" wellness. Plus, we're sold the idea that the solutions to our problems should be quick, simple, and attainable as long as we have our credit cards ready.

Anything that facilitates a stronger Self-to-part connection is healing. Compassionate understanding coming from Self can help even our most burdened parts and thus ease tension in the entire system. This isn't a quick fix, as all relationships take time. Self energy is the resource that's needed, and good news! We don't need to go looking outside ourselves for it; we just need to learn how to better access it.

Assessing Readiness and Asking for Help

Everything you've been practicing up to this point in this book has supported you in connecting with your parts from a Self-led place. If your protectors need more time to build trust in you, I recommend going through the previous chapters again. If needed, take a short, time-limited break (e.g., set an alarm to return in a few days) and try to move through the reading and exercises again. Be patient with yourself (or work with parts that are feeling impatient with your process).

Every step of the process can take a variable amount of time. To offer something that might balance expectations, I want to share what IFS therapy can look like. For some clients, it's hard to even believe that there's a difference between them (Self) and their parts, as some of the parts that lead the most feel like they *are* the whole person. We may spend several sessions, sometimes weeks, on unblending so that there can be the tiniest experience of what it's like to feel Self energy. Once a part is identified, it may take weeks to months to years before a protector is willing to let the client connect with an exile. Unburdening is only possible after a part is fully witnessed. Gentleness and consistency will be invaluable as you engage in this work.

If you feel stuck after trying to use the tools in this book, it may be time to seek outside help. The presence of a nonjudgmental therapist can help you access enough Self energy to be able to connect with your parts more easily. You could consider it "borrowing someone else's Self energy" until you can access more of your own. Being witnessed in your healing work can be powerful and provide additional healing. When it comes to the most severe experiences of trauma, abuse, or neglect in your life, having a professional who can guide the process is necessary and invaluable. Even after over two decades of therapy and my own clinical training, I still need a therapist to help me with the things that I can't face entirely on my own.

If, with some practice, you've been able to connect with protectors and that has led you to a connection with an exile, try the following steps to help parts unburden. In the next section I lay out this full process, including some of the things you've already been practicing. I'll offer an example and then guide you through your own process.

The Steps of Unburdening

The steps of helping exiles release burdens are often referred to as the healing steps. However, as I've mentioned, I believe every step of this process is a step toward healing. Before we continue, here are some disclaimers:

- Some of the things that arise may not seem logical to you. Some parts are developmentally frozen in time. These are parts that aren't concerned with logic or rationality or what's true today; they may operate in the realm of emotional or somatic experiences.

- Parts can express themselves without words, such as through images or body sensations. This is often the case for parts that experienced hurt during a preverbal stage in childhood.

- There are some prompts for imagery or envisioning these steps. Not everyone experiences their parts as visual; I encourage you to go with however you experience your parts. You aren't doing it wrong.

- Though I've laid out the process in a step-by-step fashion, it isn't 100 percent linear. You may weave back and forth between these different aspects of the unburdening process.

- There are elements of this process that may seem a little "out there" or "woo"; if you can, try to trust the process or at least give it a try (or tend to parts that feel skeptical).

Throughout the process I encourage you to notice which parts are present; if you find yourself struggling, consider there may be protectors that need additional time and attention from you.

Step 1: Ask for Protector Permission and Co-Witnessing

These initial stages of the process (often referred to as the "six Fs") were covered in the chapter 4 section titled "The Six Fs Conversation with a Protector." As discussed, you'll need to start by building a meaningful and sufficient relationship with protectors and get their consent before moving forward. Any attempt to bypass protectors and go directly to an exile will result in challenges and delays. The exercises and meditations in chapter 4 and 5 set the stage for you to do what's to follow.

Protectors who allow you to connect with an exile are still important in the process. Ask them to let you lead but **invite them to be co-witnesses** alongside you. It's healing for protectors to witness the development of a relationship between you and exiled parts; it means that they aren't responsible for taking care of the exiles all by themselves.

Step 2: Establish the Smallest Connection and Build the Container

The concept of the "smallest connection" was discussed in chapter 6. Review, practice, and hang out in this stage of the process as long as is necessary. Even if it doesn't seem like much, just making contact (e.g., sitting in silence with mutual awareness) with an exile is a giant step. While in connection with an exile, keep checking to see if you're leading with Self energy. You may have to persistently ask other parts to unblend or address their concerns along the way so that this is possible.

There are a few other things that may need to happen to set the stage for further connection with an exiled part. If the part doesn't know who you are, you may have to provide some **updating**, such as telling the part who you are, sharing what year it is, and showing the part what your life looks like now. Some parts are surprised or relieved when they learn who you are and what your life is like now. Some parts, once realizing that you aren't existing in the same time or place as they are, might become more aware of the possibility of being elsewhere. They may need help leaving a scene or moment in time that they're stuck in, sometimes called a **retrieval**. This means that we can offer the part a choice to leave where they are and be somewhere else to continue connecting with you. Ask the part and see where it wants to go (e.g., a beach, a comfy couch with dogs, with you in your heart).

Step 3: Bear Witness to Deepen the Connection

As you stay with the exile, you may notice more of a connection or sense of trust building. At this point, you can float out an invitation to this part to share whatever it would like to share with you (in words, images, sensations, or emotions). This is the heart of the witnessing process. Throughout the process you'll want to offer care and presence to the part, letting it know that what it's sharing makes sense, that you're really gaining an understanding of what it's been like for this part. You may ask the part the following questions to facilitate greater witnessing and understanding:

What would you like me to know about you?

What's it like to be you?

What was it like to go through that (experience)?

What did you need at the time?

What did you come to believe about yourself?

Did you take on any negative beliefs about yourself at that time?

Is there anything you need from me or with me?

Is there more that you want me to know?

The witnessing process may take a long time. You may have to do this a few times or many times until the part has shared everything it wants you to know. The key question in this part of the process is the question, **"Is there more?"** You can also ask the part directly: "Do you feel fully understood by me?" The witnessing process continues until the part lets you know that it feels fully understood by you and it has nothing else to share about its experience.

Step 4: Listen for Burdens

As the part shares, listen for the burdens it names either spontaneously or in response to you asking what negative beliefs it has taken on about itself. Usually these statements clearly indicate that the part has come to believe that there's something wrong with it or painful about its existence, such as:

I'm alone.

I'm unlovable.

I'm not good enough.

I'm not safe.

It's not okay to be me.

There's something wrong with me.

It's not okay to have feelings.

I'm worthless.

I don't matter.

I'm not important.

Sometimes burdens aren't in the form of beliefs or negative statements but rather sensations (e.g., cold hands), emotions (e.g., a heavy sadness), or energies or images (e.g., a swirling gray ball of pain). Whatever it is, it's something that has been painful for the part to carry or believe.

If a part opens up about its burden to you, try to remain Self-led. Notice if there are any well-meaning managers that try to reassure or tell the exile not to feel the way it does, and ask those managers to soften back. Instead, you can respond by saying things like "I'm hearing that when that happened to you, you took on the belief that you don't matter" or "I'm getting that you've been carrying around this feeling of yuckiness for a long time."

Step 5: Introduce the Possibility of Releasing the Burden

After the part has indicated that it feels understood by you and you've identified the burden, you can let the part know that if it would like and feels ready, it can release the burden in whatever way feels right for it. Let the part know that there's no pressure to do this, but see if it likes this idea and feels ready. A lot of the time, an exile who feels understood and is offered the possibility of releasing the burden will answer with a definitive "yes." If it isn't ready, see if more witnessing is needed. Or ask the part what it's worried will happen *to it or to you* if it releases the burden. Sometimes the part is ready to let go of part of the burden, not the whole thing. That's okay. You can proceed, if the part wants to, with whatever amount that the part is ready to release (e.g., 25 percent of the grief).

Step 6: Provide an Invitation and a Choice for Unburdening

Here you invite the part to release whatever portion of the burden it's ready to release. Do this by asking the part to gather up all the energies, emotions, sensations, and beliefs it doesn't want to hold on to anymore and then release it all. The part will often know what needs to happen. If it doesn't, you can offer possibilities, such as sending the burden out to the sky, to outer space, to the center of the earth, or to an element such as fire, water, earth, or air/wind, or the burden can dissolve into nothingness. You don't have to come up with anything fancy; just listen and let the part decide. Then stay with the part for as long as it needs you to while it lets go of the burden. Wait until the part lets you know it's complete, or you can ask it directly if it feels complete.

Step 7: Invite Positive Qualities and Gifts

The process of unburdening often frees up space for something else to take its place. Here we can ask the part what qualities or beliefs it might want to invite in now that there's space for it.

Sometimes the part wants something that's the opposite of the burden (e.g., a sense of connection, confidence). It can be whatever the part wants. The part can then bring in that quality or belief however it would like to. Again, the part decides.

Step 8: Integration

Now check to see how the part is doing. Once unburdened (or partially unburdened), you may notice some shifts. The part may look or feel different. Check to see that the protectors that you started with have taken note of what has happened. If they were co-witnesses to the process, they may have also shifted or released some of their burdens (e.g., responsibility). You can ask them directly if there's anything they would like to release or invite in. You can also ask them if there's something else they would like to do in your system now that the part they've been protecting is unburdened. Sometimes parts will want to keep using their skills but in a more balanced, unobligated way. They can take on certain roles without pressure and choose to rely on you to be the one to respond to situations that need attention.

Step 9: Intention

Unburdening can be very freeing for our parts, but it typically requires follow-up to anchor the changes that have occurred. You can ask parts what they might like to happen next. Would they like you to check in on them once a day? Spend time with them? Go to a playground in real life? Ask and see how the now unburdened part would like to maintain some sense of connection with you. Follow-through and consistency are key.

Step 10: Appreciation

Finally, it's important to acknowledge that this inner work has involved some amount of risk for parts in your system. Send your parts appreciation for allowing this work to happen, for trusting in you, for allowing you to know them better. Send yourself some appreciation as well for being committed to doing this healing work.

Example: Anything But Boredom

The following is an example of what an inner dialogue might look and sound like based on my own experience. Here you see this written out, but it doesn't have to be. Most of the time in this work it's completely internal; in IFS therapy, the dialogue with my parts doesn't have to be shared out loud with my therapist.

The protector here is a part that uses anything it can find to distract me and keep me from feeling bored. In getting to know this protector (firefighter), I learned that it took on this role when I was a young latchkey kid spending hours alone after school. I'll start in the middle of the process, right after the protector has given permission for me to connect with the exiled part.

> The exile is a younger version of me sitting alone in the dark, curled up in a ball, sobbing. For this part, it feels right to use they/them pronouns and I'll call them "Little Me."
>
> I let Little Me know that I'm here and notice how they respond to my presence. They look up slowly and seem surprised to see me.
>
> I ask if it's okay to get a little closer. They nod and wipe some of their tears with their sleeve. I sit down on the floor next to the part and see how they respond.
>
> Little Me: "What are you doing here?"
>
> Me: "If it's okay with you, I'd like to get to know you. Do you know who I am?"
>
> The part shakes their head no, so I let them know that I'm them, but all grown up. I show them some images of what my life looks like today. I see their jaw drop.
>
> Little Me: "I've never seen someone who looks like you before." They seem to be very interested in my very short hair, and I get a sense that they're taking in that they/I get to express our gender differently from what they're familiar with. They stare at me for a few moments, and I just stay there with them, allowing them to take in my presence.
>
> After a minute or so, they sidle up to me and lean their head on me. Since they seem to want connection, I put my arm around them.
>
> Me: "Is there anything you want to share or show me about what it's been like for you here?"

Little Me: "I don't like it. It's boring. I'm all alone."

Me: "I hear that it's boring and you're alone here. How did you get here?"

Little Me starts sending me images of what it was like for them: sitting for hours until my parents and siblings came home from work or school, waiting after school and being the last one to get picked up, trying to find things to do to stay busy. As they share with me, I feel a lot of care and compassion for them, so I send that energetically toward them and notice how they respond.

Little Me: "I don't want to be here. It's yucky."

I ask Little Me if they'd like to leave this time and place and if they want help doing that: "Would you like to come be with me somewhere else?"

Little Me: "Can we go outside?" I say yes, anywhere they'd like to go. They show me pictures of a hill overlooking the ocean.

I take this part to this new scene. Already I notice more color coming into their cheeks and their spirits seem lifted.

Me: "What else would you like me to know?"

Little Me: "Nobody cared about me."

Me: "I'm so sorry that nobody was there to give you the care you needed."

Little Me nods. They start to sob, and in my body I feel heat in my chest. I feel myself tear up, understanding that this young part of me is showing me what it felt like to be them. I allow the feelings and sensations to be felt until they start to subside.

Me: "Thank you so much for sharing what it's been like for you. I was wondering…being left alone like that, what did you start to believe about yourself?"

Little Me pauses and responds: "That no one wanted to spend time with me."

Me: "Is there something you think that means about you, that no one wanted to spend time with you?"

Little Me: "That I'm unlovable."

Me: "That makes sense you might think that if no one was around. I'm so sorry you've had to carry that around for so long. If there was a way to let go of that belief, of not being lovable, would you be interested in that?"

Little Me looks interested and nods.

Me: "Okay, then take a moment to just scan your body and gather together the belief of being unlovable and anything connected to that. Take your time."

Little Me: "Okay." There's a long pause.

Me: "When you're ready, you can let go and send away all of that. Everything you don't want to hold on to anymore."

Little Me: "I have this big ball of mud. I want to send it to the sky! Away!"

Me: "Feel free to do that. And let me know when that feels complete."

Little Me shows me that they're sending away a ball of mud to the sky until it dissolves and disappears.

Me: "Now is there anything you want to invite in? Anything you'd like to have now?"

Little Me: "I *am* lovable."

Me: "Take your time, and bring in that sense of being lovable."

Little Me: "Okay."

Me: "Is there something you'd like to do now?"

Little Me: "Can we go play? With puppies?"

Me: "Yes, we can do that!" Little Me giggles and jumps up and down. The visual imagery that comes up is a bunch of pug puppies running up the hill to where we are.

I now check to see if the protector I connected with earlier, the one that uses the strategy of distraction, has witnessed this young exile's unburdening. I ask if that protector would like a new role. The protector says that it feels relief and doesn't feel responsible anymore. It says that it wants to help me have fun but not in a way that disconnects me from the moment.

> I ask all parts if there's anything they would like moving forward, including if they would like me to check in with them. Little Me says yes, that it wants to go on walks with me and for me to play with dogs more. I say yes, I can do that.
>
> Finally, I send appreciation to all my parts for allowing me to be with them.

Keep in mind that every part will need something a little different. If I don't know what to do, I can always just ask the part and they can let me know.

Pause and reflect: What reactions or parts come up as you read through this example?

_____
_____
_____
_____
_____
_____
_____

In the following worksheet, you'll consider "The Six Fs Conversation" in chapter 4, then use the practices in chapter 5 to further support any unblending that's needed. Set aside at least twenty to forty-five minutes to engage in this guided process. A copy of this worksheet can be found at http://www.newharbinger.com/55282.

Worksheet: Guided Prompts for Unburdening

Have you made a sufficient connection with protectors, and have they given permission to work with an exile they're protecting? If not, go back to practices in chapters 4 and 5. If so, invite protectors to be co-witnesses in this process.

_____

Have you made the smallest connection to start a relationship with an exile? If not, go back to chapter 6 to establish this. If so, continue.

_____

Check for Self energy: How do you feel *toward* the part? Send whatever quality of Self energy is present. How is the part responding?

_____

_____

Updating: Does the part know who you are? If not, tell or show the part who you are. How is the part responding? (This step can happen at any time in the process, or it may not be necessary.)

_____

_____

Retrieval: Ask the part if it feels stuck in the place where it is. Offer the possibility of leaving this time and place and going wherever it would like with you. Note the part's response. (This step can happen at any time in the process, or it may not be necessary.)

_____

Witnessing: Ask the part to share its experience with you, which may include its current feelings or what it was like for them to go through whatever they did that led to being hidden away. Note what the part shares. Take however long is needed to hear from the part, offering it care and understanding throughout the process. You can ask, "Is there more you want me to know?" Don't rush through this part. Continue until the part indicates it feels fully understood by you.

_____

_____

_____

_____

_____

_____

_____

_____

_____

Burden: Ask the part what it came to believe about itself as a result of its experiences. How does the part respond? It will typically be a negative self-belief starting with "I am…"

_____

_____

_____

Unburdening: Ask the part if it's interested in letting go of the burden. If it has concerns, listen to those. If the part is interested and ready, invite it to gather everything (beliefs, feelings, energies, sensations) it wants to release and to release all of it in whatever way it would like. Note what happens.

_____

_____

_____

Invitation: Ask the part what it would like to bring in (e.g., qualities, beliefs) and invite it to do that. Note what happens.

_____

_____

Integration: Notice how the part is doing now. How has it shifted? Now check in with protectors/co-witnesses and see how they're responding to what has happened. Offer them the possibility of unburdening anything they don't need anymore or inviting in what they would like now (e.g., a new or different role). What do you notice?

_____

_____

Intention: Ask all parts what they would like from you moving forward. Would they like you to check in with them? Spend time with them? Do anything else in particular? What can you commit to doing, and do what you need to ensure you can follow up with this commitment. If you've needed to pause at some point in the process (e.g., witnessing requires more time than you have and you need to return at a later time), make a commitment regarding when you can continue.

_____

_____

_____

Appreciation: Thank any and all parts who have allowed you to do this work and know them better. Note what happens.

_____

What was this entire process like for you? What did you learn? Jot down any notes.

_____
_____
_____
_____
_____

Summing It Up

In this chapter you've learned about the process of connecting with exiles and facilitating unburdening. All of this work builds upon itself and depends on the strength of the Self-to-part relationship, and relationships take time! Now that you've learned the main components of working with parts, the next chapter will move into discussing some topics that are specific to queer and trans communities.

Chapter 8

Coming into Our Genders

Gender is a huge topic, and I imagine that you, as readers of this book, represent a wide range of identities. No matter where you are in your identity, process, and lived experience of your gender, I hope this chapter allows you to connect with whatever parts of you need to be witnessed by you. First, we'll explore the question of how parts might be related to different aspects of identity.

Are Identities Parts?

There's no easy way to define the relationship between parts and identity, and there certainly isn't a "right" way to experience your identity with regard to internal parts. Any way that you experience your gender, sexuality, or any other identity is valid.

Most people don't consider cultural identity or an otherwise integral aspect of who they are as just *one* part of them. Some identities may feel like they're connected to Self, as they're more constant throughout an entire internal system. In contrast, there are aspects of identity that may vary across the system.

For example, I being Chinese American feels integral to who I am; I don't just have a "Chinese" part. However, I do have parts that vary in terms of their relationship to the fact that

I'm a Chinese person with immigrant parents who was raised in the United States. They may have different feelings about cultural values, how I've been treated as a result of my ethnic background, or how much they're concerned with or even thinking about issues related to race or ethnicity. Your experience of your ethnic background may be vastly different from mine and just as valid.

On the other hand, like many people, my parts vary in age, some being children, some teenagers, some adults, some elders, and some that don't seem to have an age. Sometimes I meet parts within me that aren't represented as people but as animals or objects or energies, and these parts don't have an age. So, this aspect of identity (age) is expressed differently across parts within the same system.

Here's another example: Some people feel that being autistic is pervasive for them, not just a part. For the people I know who have come to embrace their autism as an expression of neurodiversity, it can be empowering to think of Self as autistic. Even so, due to society's negative reactions to neurodivergence, some parts may use strategies to mask neurodivergent or autistic ways of being. And then there may be specific parts that have characteristics that reflect some aspects of neurodivergence (e.g., some characteristics of autism) and other parts that have characteristics that reflect other aspects of neurodivergence (e.g., some characteristics of ADHD).

So the answer to the question of whether identities are parts is: *It depends.* No one else can argue with your lived experience, and that includes how you experience your internal system with regard to the identities you hold in the outside world. We also don't want to make blanket assumptions about anyone else's parts in relationship to their personal or cultural identities.

Gender Identity and Experience

Although the important identities we hold aren't simply/always parts, there's a little more nuance and complexity when it comes to gender and sexuality. There's no right or wrong; gender might be experienced differently by everyone.

To inspire deeper reflection about gender, my colleague Nic Wildes often asks the question, "Is gender a part?" First and foremost, this is a prompt for each of us to take a moment to sense into how we experience gender (or any other identity). Try to lean into what feels true for you. We don't *decide* how we experience our parts. The good news is that we can just notice, ask, and listen. We don't have to make anything up. We can try to be open and receptive to what our parts tell us about who they are and what their experiences have been.

For some people, gender identity may feel like a constant sense of who they are, perhaps even connected to their experience of Self. Even so, that person can have parts that have genders that differ from that identity. I want to emphasize that although many people, not just trans people, have parts that are a different gender from what's seen on the outside, this doesn't equate with lived experience as a trans or gender-nonconforming person who faces institutional and societal barriers and challenges in the external world. It isn't a stretch to imagine that in a patriarchal society, some cis women might have managerial or even critical parts that show up as men. This doesn't indicate being a man or having a trans or nonbinary identity (unless that particular individual indicates it). The following worksheet will help you explore gender in your own internal system. A copy of this worksheet can be found at http://www.newharbinger.com/55282.

Worksheet: Gender in the Internal System

Take a moment to reflect on your own sense of gender, however varied it may be or however much it has changed over time. Write down all the words that you've ever used to describe your gender (whether or not you've share these words with others).

_____   _____   _____

_____   _____   _____

_____   _____   _____

_____   _____   _____

_____   _____   _____

_____   _____   _____

_____   _____   _____

_____   _____   _____

_____   _____   _____

Next to each word, write whether you've experienced it as an aspect of Self, a part, both, neither, or are unsure. Then jot down any notes about why you wrote what you did.

Now note any parts that you're aware of right now (parts having reactions).

_____

_____

_____

_____

Finally, take a moment to appreciate the specific ways that these terms have allowed you to honor different aspects of your gender while also being open to the possibility that these could change even more across time. Jot down anything else you'd like.

_____

_____

_____

_____

_____

_____

_____

Parts Impacted by Societal Gendering

We've all been impacted by larger systems, including those tied to gender, such as colonialism, patriarchy, and the binary gender system. These can be parts born out of our lived experience or those that are inherited from other people in a family or culture. Some of our parts may be impacted by or have more to say about these systems than others.

It's not uncommon to have things projected onto us as soon as we're born based on our assigned/assumed sex (based on genitalia or chromosomal makeup). Sometimes others' expectations begin before we leave the womb, leaving little freedom for us to make our own choices. Some of these assumptions include: *girls wear pink, boys wear blue, girls like dolls, boys like trucks, girls are emotionally sensitive, boys don't cry*, and so on. Depending on our familial or cultural upbringing, there are many other variations of expectations based on gender.

Many people develop parts that feel stuck or powerless because they don't want to *always* follow the gender rules but know that there are consequences to breaking them. Other people may develop parts that resist and rebel. Some parts may be holding hurt because they've been punished for breaking the rules. We can have layers and layers of parts, not all known by us, that have been affected by gender socialization.

Navigating Trans Experiences

There's no universal trans experience, so there's no one right way to be trans or nonbinary. As I speak about the binary gender system, I want to make clear that there's nothing wrong if having a binary gender truly fits for you. The problem is when a binary expectation is imposed on everyone and enforced in ways that take away personal freedom.

Situations in which trans people have had to prove our right to exist or prove that we're "trans enough" to providers or other authority figures can elicit a lot of reactions from parts. Sometimes our manager parts are truly helpful when we face systems that grant or deny access to lifesaving, gender-affirming health care. Knowing about certain requirements we must meet in order to be seen as deserving of having our medical or access needs met, these managers may lead us to describe ourselves in ways that aren't totally aligned with our truths in an attempt to gain access. Sometimes these choices are Self aligned or in service of what the whole system needs, but there may still be parts that carry burdens related to not being fully witnessed or respected as we/they are.

Gatekeeping systems create barriers that are truly harmful to our mental and physical health. As a result, some parts opt for more extreme ways of coping with these realities. Firefighters' numbing behaviors may increase. It makes perfect sense that in the face of a society that's suspicious of, indifferent to, or in outright attack of who we are, our internal systems as trans people would take on more extreme burdens that call for more extreme strategies from protector parts.

To be clear: **People who are trans and/or nonbinary should not have to be in perfect mental health (or without trauma) in order to access gender-affirming medical care.** Cisgender people aren't required to have unburdened systems (or lack of mental health symptoms) in order to access medical care. While having a less burdened internal system (or even being aware of internal parts) can support going through the process of seeking and accessing medical interventions with more ease and confidence, psychotherapy shouldn't be required in order to access these services (Coleman et al., 2022).

You might be wondering: how parts relate to gender dysphoria. Gender dysphoria isn't just one thing, so there isn't just one answer. There's the more internal experience of dysphoria: distress related to one's body and gender identity not being fully aligned. But then there's the dysphoria that's more related to external systems; it may come up when having to interact with people and systems that can't see or validate who you are (e.g., being required to give/use a legal name that isn't affirming of your gender). With IFS we want to help burdened parts heal so that our protector parts don't have to work so hard using strategies that ultimately aren't helpful in the long run. For example, a protector who previously didn't let a person come out as trans might be more willing to allow someone's inner knowing to be a lived experience. An unburdened system may feel more empowered to seek whatever kind of gender affirmation is desired or needed, thus reducing dysphoria or increasing euphoria.

Some of our parts may continue to carry burdens even if we've been able to access everything we want to affirm our gender or transition (e.g., name change, surgery, hormones, accessible all-gender restrooms), depending on whether these are desired for any given person. We may feel better overall, but some part of us may hold on to a sense of not being good enough. Or, some people internalize cisnormative gender so deeply that they judge other trans people for not acting or looking masculine enough, feminine enough, trans enough, or nonbinary enough. This is an example of when our parts with burdens around gender rules can lead us to impose these same rules on others (e.g., trans peers, our own children).

The following meditation is for people of all genders. The intention is to help you get in touch with any parts that may need attention or witnessing around gendered experiences. As always, everything is an invitation, so trust whatever your system needs. An audio recording of this meditation can be found at http://www.newharbinger.com/55282.

Meditation: Gendered Experiences

1. Start by getting into a comfortable position and take some time to arrive in this very moment, wherever you are. Feel the support of the chair or seat underneath you. If you'd like, you can also envision the earth supporting you.

2. Just notice how it feels as you breathe in and breathe out, inhaling and exhaling. Notice if the quality or pace of your breath changes as you focus on it.

3. Now I invite you to turn your attention to your experience of being gendered in the world. This is a practice of just noticing, so as I guide you through this scan of your present experience, see if you can observe without staying too long on any sensation, emotion, or image.

    - What sensations do you notice in your body?

    - What feelings do you notice? Notice any pleasant or unpleasant emotions. Notice if there's any aversion, any part of you that doesn't want to think about gender. If there's an absence of feeling, just notice that.

- Notice if there are any images or memories coming to mind. Just observe whatever shows up.

4. There are probably many ways that your gendered experiences have impacted you. If you can, bring compassion to whatever parts of you have felt that impact or found ways to cope with that impact.

5. At any point if you get lost or distracted, perhaps acknowledge that these may be parts trying to protect you. You can thank these parts, and then return to your breath.

6. I'm now going to offer a series of statements. For each one, notice the feelings and reactions that come up. Allow yourself to feel whatever you feel, without judging or needing to fix or change anything. Pause as needed to acknowledge and witness the parts that are responding. Notice if the following statements are true for you:

- You were ever expected to dress in a gendered way that didn't feel right for you.
- You were ever told not to express an emotion because of a gendered expectation.
- You've ever worried that you were not soft enough or tough enough.
- You've ever been asked if you were a boy or a girl.
- You've ever not worn something you liked because you were worried it would look too feminine or too masculine.
- You've ever dieted or exercised to change your body size, shape, or how people perceive your gender.
- You've ever been told to just "act like a man" or "act like a lady."
- You've imposed gender rules on someone else, such as a partner, sibling, or child.
- You've ever felt limited in what careers were open to you based on your gender.
- You've ever been afraid to use a gender-segregated public restroom or avoided going places because of this.

- You've ever not known what to check in the "gender" box on a job application.
- You've ever felt happy when someone referred to you with a different gender pronoun than the one associated with your sex assigned at birth.
- You've ever thought it wasn't healthy or okay to be trans and/or nonbinary.
- You've made mistakes in learning how to best affirm others who are trans and/or nonbinary.
- You've experienced a sense of distress related to gender.
- You've experienced a sense of feeling affirmed in your gender.

7. These statements may have brought up different feelings, different parts of you. Let these reactions be sources of information about where to offer some attention or care. No matter what has come up for you, see if you can just let it know that it belongs…see if you can send it some compassion. We've all been affected by gender training, and we've all participated in it.

8. Send appreciation to whatever parts allowed themselves to be known or seen by you.

9. Take three more deep breaths, and when you're ready, bring your attention outward again.

10. Jot down notes or journal about what came up for you or what you want to return to. Do whatever feels like the most caring thing for you and your parts.

Pause and reflect: What came up for you during the meditation?

_____
_____
_____
_____

Unburdening gender is no simple or small feat. Some of us know in our heart of hearts who we are despite the world being unable to witness or respect us. This deep wisdom is an aspect of our own Self energy. In practicing this next meditation, you may access a greater capacity to tune in to that Self energy and hold space for the parts of you that carry gender-related burdens. You may have a completely different experience each time you engage in this practice. An audio recording of this meditation can be found at http://www.newharbinger.com/55282.

Meditation: Unburdening Gender, Honoring Our Truths

1. Begin by taking some deep breaths. Let your internal system know that you'd like to do some focused work with parts that are holding pain around gender.

2. If you can, allow any part that has something to say to have a little time to share and be acknowledged. You might even imagine them passing a mic along, sharing whatever is most important to them. In this practice, it's important for all your parts to know that there's space for them.

3. When each part has had a chance to speak, ask the whole group: Who needs the most attention today? See if there's a part that steps forward, or listen for any wisdom coming from the core of who you are. Ask the rest of your parts to give you space to be with and learn more about the centered part.

4. Allow yourself to focus on the centered part, however you experience it. Are you open to hearing from it?

    - If not, spend some time with whatever part might have concerns, letting them know that your intention is to help each and every part of you, but one at a time. Then see if it might be willing to give some space, even a few minutes.

    - If the part is willing to give that space, thank it and return to the centered part. If it isn't, then this part may be the one that needs the attention right now. It may become the new centered part. Whatever part you're with, just see if there's an openness to spending time with it.

5. Invite the centered part to share anything it wants you to know about what it has experienced or what it has been carrying. Try to offer full presence to this part and perhaps send care or compassion if that feels available.

6. Notice how the part responds to you. Can you sense a connection being built between you and this part? You might ask the part if there's anything that it would like from you. Perhaps it would like to share more, or maybe it would like for you to just sit with it, or perhaps it would even like you to give it a hug or hold it. For parts who have never had much space to be seen or heard, even making contact and learning a little about them can be a really big step.

7. Some parts aren't aware of the present day or who you are today, so if needed, you can update this part on what's true today.

8. When it feels right, ask the part if it feels understood by you. If not, ask what else it needs you to know. Then ask the part what beliefs it has taken on about itself or gender as a result of these experiences and feelings. The part may share some of the burden it has been carrying with you. Let the part know that it makes sense it took on that belief or energy at the time it did, and then let it know that there's a possibility it could at some point release what it's been carrying. See how it responds.

9. If the part seems interested, you could ask the part if it feels ready to release any amount, even 1 percent, of the burden it's been carrying. If the part feels ready to let go of any amount of the burden, invite it to do that in any way that feels right for it in this moment. Keep offering presence to the part as it releases whatever it would like to.

10. After that happens, you can offer the part any or all of these options: The part may want to invite in positive qualities or energies now that some space is freed up, perhaps a way to relate to gender that feels more aligned. The part may ask you to come back and spend a little time with it throughout the coming week. Or the part may ask you to return to it at a later time so that you can learn more and perhaps it might be able to continue the process of releasing its burdens. See whatever else the part would like.

11. Finally, send appreciation to this part and every other part for this chance to connect with them. Then take a few deep breaths and bring your focus outward again.

12. You may decide you want to repeat this exercise but with a different part. You may want to jot down some notes or journal about what you learned or any intentions to follow up with any of your parts.

Pause and reflect: What came up for you during the meditation?

_____
_____
_____
_____
_____
_____
_____

Summing It Up

In this chapter you've explored how you've experienced your own gender. You've gotten to connect with some of the parts who have been deeply affected by gender norms and expectations. If you're questioning gender or what it means to be trans and/or nonbinary, I recommend the book *Am I Trans Enough?* by Alo Johnston, and *decolonizing trans/gender 101* by b. binaohan. In the next chapter, we'll look at the ways that aspects of sexuality and relationships relate to the internal system.

Chapter 9

Sexuality and Relationships

In this chapter, we'll explore the different facets of how we, as queer people, interact with and relate to others and society. We'll look at sexual attraction, forming relationships, attachment, and navigating cultural burdens related to sexuality and relationships.

Sexual Attraction and Orientation

There are as many ways to relate to sexuality as there are people. We may consider our sexuality to be a constant part of who we are, connected to our Self energy. Or some of us may experience our sexuality as parts of us with different attractions or inclinations. To provide some examples of the kinds of diversity that can be present, here is a list of the ways that parts might relate to or express sexuality:

- Some parts may be attracted to some bodies or gender expressions, while other parts may be attracted to other kinds of bodies or gender expressions.

- Some parts may be very interested in sex, and others may be less interested or not interested at all.

- Some parts may be more drawn to kink/BDSM, while others are not.

- Some parts may be more concerned with romantic attraction, while others may be more interested in sexual attraction.

- Some parts may be monogamously inclined, while others may be polyamorous.

- Sometimes there may simply be an identity as queer (perhaps Self is queer!), without any parts that have strictly defined or constant attractions.

These are just a few examples. Beyond parts that express themselves around facets of (or identities based on) sexuality or relationships, you may have parts that have feelings about or have taken on burdens related to these realms.

Navigating Expectations Based on Sexuality

Living in a heterocentric (and homophobic) world, we're taught that we should be attracted to people of the "opposite" binary gender, get married, and never question this path. As queer and trans people, we know that these rules don't work for everyone. Socialization around sexuality and relationships is deep, and most people carry some "baggage" (i.e., burdens) related to sex, sexual attraction, and relationships. The messages we receive can be rooted in family, cultural, and sometimes religious norms. It's not surprising, then, that we may have parts with reactions to societal norms *or* the consequences of violating these norms.

Some people grow up with queer family members, but the vast majority of us have been "the only one" in our family. Having an awareness of being different from other family members (e.g., with regard to sexual orientation) when other aspects of identity are the same (e.g., access to financial resources) can take a toll. Our parts may take on burdens of aloneness, of not being acceptable, or of not being seen. Even if our family members try to be allies, they don't always know how to put their support into action. This can create real barriers to feeling understood and connected.

Even if we feel confident in our queerness and are "out" to others, our parts may still be burdened by heteronormative expectations (sometimes even within queer relationships). Or we may feel that being accepted in queer community comes at the expense of exiling other aspects of who we are. Many visible queer and LGBTQ+ communities are white-centering, so this can create a sense of feeing included and excluded at the same time for BIPOC queer and trans people. Our parts may feel burdened by having to choose one identity over another when both/all are important to us.

Unfortunately, many queer people have experienced outright rejection or exclusion, resulting in parts that carry burdens of shame or feeling damaged, unsafe, or unlovable. In some cases, our protectors make sure we hide our truths and never come out; this is often with the positive intention of maintaining safety or belonging. Sometimes these protector parts take on strategies of conforming and assimilating at any cost; they may lead to choosing partners or relationships that aren't in line with our deeper needs or desires.

Bisexual people often face biphobia within queer communities; burdens of invisibility or not being believed are common. And then there are people who know themselves to be queer but are asexual or aromantic. And those people may not feel completely welcomed in queer spaces that assume everyone is allosexual. It's strange that people who've experienced exclusion due to sexuality can perpetuate that exclusion. But if we view that from a parts lens, it makes sense. Burdened parts can perpetuate the harm they've experienced.

Consider this example. Sun (he/him) grew up in a religious community that viewed sex as only appropriate between a (cis) man and a (cis) woman within a monogamous marriage. Aware of his attractions to men at a young age, a young part (exile) took on the burden of shame and the belief that "there's something wrong with me." In response, Sun's manager parts made him follow the expected path of marrying a woman. His firefighter parts pushed his authentic feelings and attractions under the surface by taking on the behavior of compulsive hairpulling (trichotillomania). Later in life, after getting a divorce, Sun sought therapy for the first time. In this process, he uncovered the parts and burdens that had been hidden away. His trichotillomania lessened as his protector softened, allowing him to build a relationship with his younger parts. Over time, healing from burdens allowed Sun to access the courage and confidence to come out and attend a support group with other gay and bisexual men who were coming out later in life.

Perhaps you can relate to some of Sun's experience; the following meditation will help you get in touch with any parts that may need attention or witnessing related to being socialized in a heterocentric and homophobic culture. As always, everything in this meditation is an invitation, so trust your system and whatever it needs. An audio recording of this meditation can be found at http://www.newharbinger.com/55282.

Meditation: Impacts of Heterocentrism

1. Start by getting into a comfortable position, feeling the support of the chair or the earth beneath you.

2. Become aware of your breathing. Notice if the quality or pace of your breathing changes as you rest your attention on it. Just notice how it feels as you breathe in and breathe out, inhaling and exhaling.

3. Now I invite you to take a moment to acknowledge the parts of you that are aware of your sexuality. See if you can just notice whatever comes up, whether it be images, sensations, feelings, or thoughts.

   - What sensations do you notice in your body?
   - What feelings do you notice?
   - What images do you notice?
   - What internal messages are present? Just notice. Right now there's nothing to do, nothing to fix. We're just noticing.

4. Notice how you're feeling toward whatever parts are present. If you can, bring care or compassion to whatever parts of you have felt impacted by expectations around sexuality.

5. Notice how your parts are responding to being witnessed by you.

6. Now see if you can get curious about what these parts might want to share.

   - How have your parts experienced expectations about sexuality from your family, peers, or society?
   - What has it been like for them to be queer in a heterocentric world?
   - Which parts have hidden?
   - Which parts have come in to help you cope?
   - If there are painful feelings that arise, that makes sense. You can let your parts know that it makes sense that they've felt hurt. And let them know that you're here, and you want to build a deeper relationship with them, one in which they feel protected by you.

7. Check to see if your parts are aware of your life today. If any of them are stuck in any moment in the past, see if you can share with them or show them what things are like now. And see how they react.

8. Take another few moments to be with these parts of you that have been impacted by heterocentrism and homophobia. Ask if they would like more time with you, if they're open to you connecting with them again to learn more about them.

9. Finally, send appreciation to whatever parts allowed themselves to be known or seen by you.

10. Take three more deep breaths, and when you're ready, open your eyes and bring your attention outward again.

11. Jot down anything that came up for you or what you want to return to.

Pause and reflect: What came up for you during the meditation?

_____
_____
_____
_____
_____
_____

Navigating Sex

There's no one way to think, talk about, or engage when it comes to sex. We all have a different relationship to sex, and we may have different internal parts that have experiences related to sex. As queer and trans people, many of us haven't had a lot of support in exploring what kinds of sex are right for us. If we learned about sex in schools, typically queer sex was not discussed. That lack of representation itself can create parts that feel confused or ashamed, or internalize that there's something wrong with us.

Having a healthy or positive relationship with sex typically means having some awareness of what we want, an ability to communicate that to anyone we choose to be sexual with, and the capacity to set limits and boundaries. Some of the parts of us that are burdened with regard to sex may lead to our protector parts choosing strategies that get in the way of feeling good or empowered in our sex lives.

Pause and reflect: Which parts of you need to be heard regarding sex? Can you take some time to check in with your system and note any parts that need further attention?

_____

_____

_____

_____

Sexual Trauma

An unfortunate reality is that being a survivor of sexual assault or trauma isn't uncommon in queer and trans communities. Trauma doesn't cause a queer or trans identity; unaddressed trauma can actually get in the way of feeling safe to be ourselves as queer or trans people because it would put us at even greater of risk of societal stigma. When sexual trauma occurs at any age, but especially for young people, it's understandable that parts respond by taking on protective roles.

There's no one way that parts respond to traumatic sexual events. Some parts might show up as trauma or classic PTSD symptoms, such as hypervigilance, flashbacks, nightmares, or dissociation. Some parts may seek out more sexual experiences, while others may avoid sex and dating altogether. Other parts may find other ways to numb feelings when the pain of the trauma (in many cases the exile's burden) comes to the surface. Sometimes within new relationships or with new sex partners, positive feelings make it possible for protectors to relax and not fiercely protect the system. But for many people with trauma, as relationships get emotionally closer, the trauma symptoms may emerge, possibly because there's more at stake (e.g., more fear of losing something that's good). In their own ways, all of these parts are operating under the tenet of "never again," meaning they will do whatever it takes to protect you from being hurt again.

It's common for the more vulnerable or traumatized exiles to take on negative beliefs, such as being at fault, damaged, or ashamed of not feeling empowered to stop the harm from happening. Young parts who are unaware of the systems of violence (e.g., rape culture, the high prevalence of childhood sexual abuse, the high incidence of intimate partner violence) are especially vulnerable to blaming themselves or feeling alone. So much of the time when young people tell someone about sexual abuse, they aren't believed. It's then easy for parts to take on beliefs about being unimportant or it not being safe to speak up. If you have these parts, know that you aren't alone. And know that for these parts, it can be incredibly healing to feel seen, heard, understood, and cared for by you. If you find that you have a hard time accessing compassion for these parts, you'll have to start with the protectors that you're likely blended with and hear their concerns.

If you've experienced sexual trauma, some of the exercises in this workbook may help you better understand what happens for you when you get triggered or have intense feelings or behaviors with regard to sex. You may be able to cultivate a greater capacity to offer compassion to every part of you that's working hard to protect you, as well as the more vulnerable parts that have lived through traumatic experiences. However, you don't need to do it alone. You may want to seek out professional support from a queer-affirming therapist who's versed in working with sexual abuse or assault.

Navigating Relational Structures and Styles

One of the beautiful things about being queer is that once we start to realize that the dominant culture's norms for sexual attraction don't work for us, we're more inclined to question other imposed norms for how we relate to and form relationships with others. Oftentimes this leads to many more possibilities and an orientation of choice rather than obligation. There's no one right way to be in a relationship; what's more important is that we get to choose whatever works best for us. Monogamy works well for some people but not others. Some of us choose to practice ethical or consensual nonmonogamy, which is a broad category that offers many options.

The ways we navigate relational structures and styles may involve a dynamic interplay of parts. For example, some people may consider themselves polyamorous as a consistent identity, perhaps with an entire system that's polyamorous. Other people may feel more fluid or have more context-dependent polyamory if they have some parts that are very much polyamorous and other parts that prefer monogamy. Sometimes polyamory is a Self-led choice that allows us to satisfy the needs of different parts of us, perhaps parts that vary in attraction to people of different genders or personality styles. Rather than trying to choose the right or wrong way to be in relationships, we can strive to be Self-led in whatever choices we make.

It may not always be an easy process to claim the right to do things differently. We may have to work with parts that feel shame about our desires, fear hurting our partners if we want something different than or in addition to what they offer us, or feel deeply insecure or jealous when our partners have other romantic or sexual interests. It's not for everyone. And, if there's an exploration that feels important for you, it will require working with parts that not everyone speaks openly about. It's important that even in the face of systems that aren't affirming or inclusive, all of your parts are welcomed by you.

For example, Ximena realized as a young adult that monogamy just wasn't right for her, but her manager parts pushed her to be in monogamous relationships in order to fit in. These parts told her that it was shameful to have attractions to more than one person (e.g., "there's something wrong with me"). Still, Ximena knew at some level that she was capable of holding strong romantic feelings for more than one person, but she kept choosing partners who were monogamous. After getting support from a therapist, Ximena chose to come out as polyamorous. On one hand, she felt free. On the other hand, there was still a young part of her who felt shame.

Doing parts work allowed her to connect to that young part. What she identified was not only shame regarding her polyamorous nature but also the ancestral burden of shame passed down from her mother's side of the family. Ximena's grandmother had been shamed in her community for having an "illegitimate" (out of wedlock) child on her own. As a reaction, Ximena's mother heavily emphasized that the right way to be in relationship was to choose one person and commit to them for life. Ximena was simultaneously carrying ancestral burdens (related to her grandmother) as well as cultural burdens (related to the doctrine of monogamy). After releasing these burdens, Ximena felt more confident as a polyamorous person. She not only started to date other people who had a strong inclination toward nonmonogamy, but she also built social community with other polyamorous people.

There's no "right" way to structure a relationship, but you may want to consider the parts of you who do think there's a more desirable way to be in relationship. Regardless of what you choose or what feels right for you, you may have parts that have reactions. Here are some examples of parts that might show up regarding relational structure:

- Parts that enjoy monogamy

- Parts that enjoy polyamory or nonmonogamy

- Parts that judge oneself or others for their relational structure (e.g., for being monogamous, for being polyamorous)

- Parts that get jealous or insecure, possibly leading to rigid rules

- Parts that struggle with setting boundaries

- Parts that impulsively break nonmonogamy agreements

- Parts that feel overwhelmed with dealing with so many people's schedules and needs

- Parts that overextend and have a hard time observing capacity, sometimes leading to involvement with too many people (poly-*over*saturated)

- Parts that experience compersion (happiness for a partner getting to connect with someone else)

- Parts that are interested in people of one gender while other parts are interested in people of a very different gender

- Parts that love being in shared community with metamours (partners' other partners)

Pause and reflect: Do you relate to any of these? What trailheads have you identified? Which parts might you need to spend more time with?

_____

_____

_____

_____

_____

_____

Navigating Attachments in Relationships

Our capacity to form close bonds with others is often related to our histories of relating to our earliest caregivers. Whether we had people who were stable or consistently present for us can determine how easy it is for us to trust our connections with others. There's an enormous body of literature on the topic of attachment and attachment styles, and in this book we won't get into great detail. But it's worth mentioning that there are different attachment styles, such as secure attachment, anxious or ambivalent attachment, avoidant or dismissive attachment, and disorganized attachment, and there are subsets of these styles. Whether or not these descriptors fit for you, you likely have parts that are involved in how you attach (or avoid attaching) to others.

In addition, attachment is often referred to as one consistent style for a person. However, different contexts may draw out different kinds of behaviors or styles. It's possible to have some parts that are more securely attached while others are more insecurely attached. Rather than work within the model of having one attachment style, you may want to get curious about specific parts and their behaviors. Do some parts get jealous, insecure, or clingy? Do some parts avoid any kind of relationship conflict and freeze up or shut down? Do you have one part that desperately seeks attention from a potential date and then another part that jumps in to make you avoid that person altogether as soon as they get too close? These behaviors are important to track, but what's more important is the underlying beliefs or burdens that the more vulnerable parts may be carrying. Believe it or not, the majority of the time these parts are all trying to maintain connection (even if it doesn't look like it on the surface). The same external behavior (e.g., avoiding conflict) may be connected to any number of burdens around connection. The following worksheet can help you get in touch with parts engaged in attachment. A copy of this worksheet can be found at http://www.newharbinger.com/55282.

Worksheet: Understanding an Attachment Strategy

Reflect on one particular strategy for navigating attachment.

Choose one specific strategy or behavior, such as shutting down when there's conflict or going into a critical/attack mode when a partner isn't available. Write that down here.

_____

Now reflect on when this behavior occurs. Can you think of any triggers or external circumstances that might draw this part into enacting this behavior?

_____

_____

_____

See if you can locate this part inside you. What images, feelings, sensations, or messages do you notice?

_____

_____

_____

Now notice how you feel toward this part, which we'll consider the centered part. If you notice judgment, ask the part with judgment to give you a little space so you can stay with this exploration. You may need to stay with other protector parts until they're willing to unblend. Then turn your attention back to the centered part. If you can access any qualities of Self energy (e.g., care, compassion), see if you can send that toward this part. And then notice: how does the part respond?

_____

_____

Now invite the part to tell you what, if anything, it wants you to know about why it does what it does. Listen. And then ask the part what it's worried would happen if it didn't do what it does.

_____

_____

_____

_____

As you get to know this part, keep noticing how you're feeling toward it. Does it make sense to you that this part is doing what it's doing? Can you access any appreciation for this part for trying to help you? If so, send that appreciation. Then ask the part if it would be willing to allow you to witness the part it's protecting. Note what happens.

_____

_____

_____

_____

If you're able to connect with and witness the more vulnerable part underneath the protector, spend some time with it. See what happens when you offer presence, care, and listening to this part. Know that the more time you're able to spend with this part, including helping it heal/unburden, the less the protector will feel compelled to do what it does.

_____

_____

Finally, you can thank these parts for helping you understand them better. If the centered/protector part shows up in a future interaction, imagine how you might respond to it. What do you imagine is possible if you could notice this behavior as a protector part and have a conversation with it like you did here?

_____

_____

_____

_____

Jot down anything else you noticed, or feel free to try this exercise again with a different protector strategy.

_____

_____

_____

_____

_____

Shifting Roles in Relationships

With societal conditioning, the concept of being "assigned" to something at birth extends beyond our genders (e.g., male, female). Gender expression can also be assigned (typically as masculine or feminine), and so can our gender roles (e.g., being the breadwinner, taking care of domestic labor). We're typically assigned a sexual orientation (i.e., heterosexual) and certain roles in our relationships and communities.

For many of us who are queer and/or trans, we literally fail the assignment. Our young or vulnerable parts are susceptible to the belief (burden) that we've failed, rather than it being a failure of the environment to support us in the fullness of who we are. If we consciously choose how to relate to the people in our lives, we may choose roles that are very different from either what was expected or what felt right in the past. For example, someone who identifies as a mother prior to transition may opt to identify as a father or a "mapa" after transitioning. A non-binary person who was previously thought to be a "brother" might opt to be referred to as a "sibling."

Even if you're cisgender, you may have roles that shift for you. You may notice that there's a certain meaning to being the gay uncle rather than an uncle who's heterosexual. Some parts of us may be aware of role shifts, while others may be stuck in a past arrangement. In doing work with our parts, we can help them all get updated.

Pause and reflect: Think of all the different roles you've played in relationship to others (e.g., sister, father, girlfriend, uncle, etc.). Which of these roles still feel aligned for you today? Which roles no longer fit? If you find that some roles still fit in some ways but not others, note that. Then note any reactions from your parts.

_____

_____

_____

_____

_____

Summing It Up

In this chapter you've gotten to increase your awareness of and connection with parts that have been impacted by heterocentric and homophobic norms for sexuality and relationships. You've also gotten to reflect on and connect with parts with concerns related to relational structures, sex, attachment, and shifting roles in relationships. In the next chapter, you'll learn more about the relationship between the body and your internal system.

Chapter 10

Relating to Our Bodies

Having a body isn't always easy. Adjusting to your body as it changes across different life stages is bound to bring up feelings. In this chapter, we'll discuss some of the context for how we learn to relate to our bodies and what it might mean to move toward greater body liberation. We'll also explore what it means to take care of ourselves and develop more of a Self-led way of relating to our bodies.

Burdened Parts, Burdened Societies

As young people, whether explicitly or implicitly, we're taught that there's a "normal" for how to have a body. That definition of normal is defined by white supremacy, colonialism, ableism, anti-fatness, cisheteropatriarchy, and the many expressions of how these systems interact with each other. It can feel impossible to live in a society like this without taking on burdens related to our bodies.

Although we, as individuals, may take on burdens as a result of systems of oppression, the larger culture itself is also burdened. Systems change is needed, *and* while these systems are still very much impacting us, we all need to find ways to care for ourselves and each other. We can ask ourselves what choices we have even though there are many things we can't change. One of the choices we *can* make is to make peace with ourselves. I truly believe that doing so creates more freedom and capacity to show up for liberation work. It's all connected.

Toward Body Liberation

While many of us have been harmed by external systems, we (or rather, our parts) have also been complicit. When parts of us internalize certain beliefs about the "right" way to have a body, the impact is amplified and it's possible that we'll perpetuate the very systems that harm us. If we want to be free as queer and trans people, we must work toward freedom for all people, all bodies. We can't just work on our relationships to our own bodies without examining the beliefs that our parts have taken on regarding *all* bodies. Before we get into how to work with our parts, here are some definitions and framings as a foundation:

- **Body liberation** is a broad movement away from the structures, policies, behaviors, beliefs, and ways of relating that are based on a "right" or "normal" way to have a body (however that's defined within a given group or culture). I recommend the book *Decolonizing Wellness: A QTBIPOC-Centered Guide to Escape the Diet Trap, Heal Your Self-Image, and Achieve Body Liberation* by Dalia Kinsey as an excellent resource geared toward BIPOC queer and trans people.

- **Body positivity** is a movement that tells us we should love our bodies as they are. Although well meaning, this approach is oversimplified and contains an agenda that's similar to toxic positivity. We can't simply love ourselves or our bodies just because someone (or a manager part) tells us to. If we can access self-love, it must come through a relationship between Self and parts. Body positivity can also be weaponized against trans people. We can't just will ourselves into loving our bodies as they are; sometimes accessing medical interventions that change the body is the most loving thing we can offer ourselves. If a part of us tells us we *should* love our body and we just can't figure out how to do that, it can result in another part of us feeling shame.

- **Body Trust** is our innate capacity to *listen* to the wisdom of our bodies. We all have this wisdom, but most of us have been trained to tune it out in order to fit in or follow rules. Body Trust is very much related to trust in Self leadership. Unfortunately, many of us, because of the expectations of grind culture, have adopted a top-down approach to having a body rather than one that starts with listening to what our body's actual needs are. Some examples of respecting our body are sleeping when we're tired, resting when we've reached our physical limits (or ideally before!), and eating what tastes and feels good to us. If you'd like to learn more about Body Trust, I recommend the book *Reclaiming Body Trust: A Path to Healing & Liberation* by Hilary Kinavey and Dana Sturtevant.

- **Healthism** is the belief that each of us has a moral responsibility to be healthy through our day-to-day choices and habits. It's a form of ableism that's widely accepted in our society. If someone isn't healthy, they're typically judged and blamed. Healthism is highly individualistic and neglects to consider the social determinants of health. It's related to wellness culture, which is the societal pressure to be "well" in myriad ways, whether it be through diet, exercise, supplements, or whatever trendy quick fixes are being sold to us.

- **Body appraisal and surveillance** are processes that involve evaluating our bodies (or others' bodies) as good, bad, acceptable, or unacceptable based on certain societal standards. Body appraisal is more related to how we assign positive or negative qualities to a body, whereas surveillance is more about the checking or monitoring of our body or appearance. It's important to highlight these concepts in order to get away from the idea that the issues we have about our body is simply about an internal "body image" without the larger context of a burdened society.

Discernment: Acceptance and Change

We all have feelings about the way we look. We aren't required to love every aspect of our body; however, we may be able to cultivate more peace with it. Acceptance of our current reality can include accepting when there's a need for change. Of course, some aspects of our bodies are changeable and others are not. It can be confusing. For example, it can be hard to discern whether negative feelings about the shapes of our bodies is related to a sense of gender dysphoria, societal anti-fatness, both, or something different altogether. Befriending the parts of us that have reactions about our bodies can help us with this discernment; if we listen to their concerns, we can better determine what might need to change versus what we'll want to make peace with.

Given our personal experiences of body appraisal and surveillance, it's easy to develop parts that engage in constant checking, evaluation, or criticism of our bodies. This can create a great amount of stress related to navigating the everyday reality of having a body. It can be helpful to connect with parts that are involved in these protective strategies.

The following worksheet will provide an opportunity to get to know a part of you that's critical of your body. The goal of this practice is to invite this part to unblend so that you can get a better understanding of its true hopes and intentions for you. A copy of this worksheet can be found at http://www.newharbinger.com/55282.

Worksheet: Honest Conversations with Body Critics

Choose a part that can be self-critical. See if you can notice whether this part seems to be present. Note how you experience it in or around your body.

_____

_____

_____

Focus on this part even more. What else do you notice about it?

_____

_____

_____

Check to see how you're feeling toward the part: positive, negative, neutral? Notice if any qualities of Self energy are present. If so, try to send that quality (e.g., compassion, connectedness) to that part and see how it responds to you. What do you notice?

_____

_____

Ask the part to tell you about itself, its role, and its intentions. You may want to include the following questions. Write down what you learn from the part.

- What do you see as your job? How are you trying to help me?
- What positive intentions do you have for me? What is it that you truly want for me?
- How or where did you learn to do your job? How did you decide what was good or bad in terms of my body?

- What would need to happen for you to have a more balanced or relaxed role? If something needs to change, do you have a sense of whether that change needs to be internal or external?
- What are your fears about what might happen if you don't do what you're doing?
- If there were a way for us to be in more conversation or collaboration, would you feel the need to be so critical? Would it be possible to tell me what you want for me without being so harsh?

- _____
- _____
- _____
- _____
- _____

Thank the part for sharing with you, setting intentions, or making agreements to be in more conversation with the part, if desired or needed. What kind of follow-up might be possible in order to stay connected with this part? Jot down any notes or anything you'd like to come back to.

_____
_____
_____
_____
_____

Self-Care and Self-Love: Not for Sale

Capitalism will tell us that self-care and self-love can be ours—if we're willing to pay the price. We've all been told that if we just (fill in the blank), we'll be happy, healthy, and desirable. The result is that we develop protector parts that buy into all this messaging and marketing, profiting from the deeper wounds and insecurities of our exiled parts. Instead of engaging within to help these hurt parts, our protectors seek external solutions that only provide short-term proxies for feeling good about ourselves and our bodies.

Unfortunately, the concept of self-care has been co-opted by healthism and wellness culture, leading to a kind of shaming or blaming if we don't engage in self-care. *Have you gotten your antioxidants? How many bubble baths have you taken? Do you need to cut out a whole food group?* These are the voices of wellness culture that our protectors can adopt as strategies to help us feel like we're doing something right with our bodies.

The good news is, working with our parts helps us tune out these external messages and tune in to what our bodies really need. This is a process that takes time. It starts with unlearning societal rules about how to relate to our bodies (working with our protectors). Then we can move toward reclaiming self-care as something that no one else can decide for us (building inner connection so that we can listen when our parts tell us what they need). Unencumbered by the moral dictates of wellness culture, self-care can simply be the actions that we take to tend to our own needs, whether they be physical, emotional, or spiritual. Self-care may look different for each of us, and what's most helpful or supportive may vary from day to day or moment to moment. We don't have to figure it out; ideally, we can open ourselves to listening or noticing what our system needs.

For many neurodivergent folks, it can be a little more (or very) challenging to get in touch with certain cues or signals that tell us what the next indicated action is. I encourage you to be patient with yourself. Sometimes a part, noticing that we don't experience the same signals as others, tells us it's impossible. Your body may have a very different way of communicating with you than it does for neurotypical people. For example, for some people they simply know they're hungry because they feel a rumble in their belly. If you're neurodivergent, you may have a more concrete, behavioral, or external cue, like noticing that a part that fidgets is tapped out to the point that a fidget toy isn't helping as it usually does. Another example is if you're someone who has a hyperverbal part, you may notice that this part is less active (as is the case for me)! And then sometimes if it's impossible to tell just by trying to check in with the body, you might need to take a bite of food to see how your system responds, or in some cases you may need to set an alarm to eat at regular intervals (e.g., at least every three or four hours).

Some of us have gotten so disconnected from our bodies that we need the guidance of professionals to help us reattune. Sometimes our disconnection from our bodies gets severe or even life-threatening, such as the case when we have a protector that engages in extreme food restriction. Just know that while the process of recovering a connection to our bodies may seem daunting, it's possible for each and every one of us.

You might be wondering how IFS can help you better attend to your physical needs. Even without explicitly putting a spotlight on self-care and the body, IFS work inevitably leads to listening to what our parts are needing. It's just that there are so many aspects of life that we have to deal with that sometimes we can forget that we even have a body!

There are different categories that each of us can attune to. There may be additional categories that you know are crucial for you in maintaining balance. The following worksheet will help you identify these categories for yourself so that you can then use IFS to become more attuned to your body and self-care needs. A copy of this worksheet can be found at http://www.newharbinger.com/55282.

Worksheet: Reclaiming Self-Led Self-Care

Reflect on these different aspects of your physical experience. You can complete this chart at any time, especially when your physical state/needs change. At the end, add other categories that are important to you. Leave the fifth column blank for now. The information gathered here will help you identify which parts you need to spend more time with.

Day/Date: _____

Self-care category	How easy is it to take action to tend to these needs? Scale of 0–10	Which parts or strategies might create challenges?	What other parts have reactions or might not want me to look more closely?	Takeaways from conversations with parts
Sleep				
Physical rest/relaxation				
Hunger/food				
Temperature regulation				
Sensory rest (i.e., low stimulation)				
Executive function rest				
Physical movement				

Creativity	Social interaction	Alone time	Play/fun or stimulating activity	Sex	Pain regulation	Other:	Other:	Other:

Review what you've written. Let's keep it simple by just choosing one self-care category that you'd like to work with. I recommend going with something that has a low to medium emotional charge for you and leaving the harder stuff for later after you've built more trust with your system. Looking at the one category you've chosen, select one part to center first. If you have something written in the fourth column, you'll probably need to start there before working directly with the parts in the third column.

Category: _____

Notice how you feel toward the centered part. Ask any parts that are jumping in with commentary to move back. If you can access a quality of Self energy, send that toward the centered part and notice how it responds to you. Stay in this part of the process for as long as is needed. Once you're able to sense a connection or at least co-awareness, proceed.

Ask the part the following questions:

- How are you trying to help me? What are your intentions?
- What's your understanding of what happens as a result of you doing what you do? What are the positive effects? What are the negative effects?
- Do you like the role you are in? Or is there something you'd rather be doing?
- What are you worried would happen to me if you didn't do the thing you do?
- What would need to happen for you to relax and find a more balanced way of doing your job?
- If I could offer attention and care to any other parts that you're protecting, would you be willing to allow the space for me to do that?

Jot down any notes about what came up in the conversation or any follow-up you'd like to do. This could include doing something the part has requested or setting aside time to connect with any younger or more vulnerable parts/exiles that need attention.

_____

_____

_____

_____

_____

Return to the final column of the chart and jot down any new insights or takeaways from this exercise.

Repeat this as needed with different categories, using this chart as a guide. You may want to repeat this exercise as your needs shift over time.

Because we're often dealing with a complex combination of factors or pressures in our everyday lives, it can be helpful to cultivate a simple "check-in" practice that we can use in any situation. The following short meditation can be used as an everyday practice first thing in the morning, throughout the day, or before you go to bed. An audio recording of this meditation can be found at http://www.newharbinger.com/55282.

Meditation: Pausing to Check In

1. Take a moment to pause. Notice your breath.

2. Invite your parts to guide you in letting you know what you need to best care for yourself and for them.

    - First start with the **body**. Do a scan of your body and notice any sensations or lack of sensations. What does your body tell you about what might be appropriate for you at this very moment?

    - Next move to **emotions**. Do a scan for any feelings or lack of feelings that are present. What do these feelings tell you about what might be needed at this very moment?

    - Next move to reflecting on **what's happening externally**. Do a scan of behaviors or external conditions that are noteworthy. Is there any new information about what might be needed at this very moment?

3. Thank your parts for allowing you to connect with them and for offering their guidance.

4. From the information you gathered from this conversation with your parts, identify the **next best action** to take so that you can move toward more physical and/or emotional balance.

Pause and reflect: What came up for you during the meditation?

_____
_____
_____
_____

Summing It Up

In this chapter we've looked at factors that affect how we experience or treat our bodies. From spending time identifying parts of us that have judgments about our bodies to the actions we take on a day-to-day basis to maintain our bodies, ideally this chapter has guided you to build even 1 percent more Self leadership in how you relate to your physical and emotional needs. In the next chapter, we'll spend time identifying our strengths as queer and trans people and communities.

Chapter 11

Recognizing Our Strengths

The challenges we face as queer and trans people may be so impactful that we forget we have strengths, resilience, and power, especially in community and solidarity. In this chapter we'll explore examples of individual and community resilience. You'll be guided in a practice of turning your attention to the parts of you that have led to your survival and strength in the world.

All Parts Are Resilient

It's not just our Self energy that provides resilience. Every single part of us is resilient. Every single part of us has unique skills, talents, and gifts. Our protectors are truly creative in how they take on their roles. Even though the outcome doesn't always turn out so well for the entire system and our lived experience, our parts are incredibly dedicated to protecting us.

Our manager parts are the ones that bring persistence and accountability. When in a balanced, relaxed, or trusting state, they pay attention to what's important to us. They help us attune to others and express ourselves. They help us track our commitments and resources.

Our firefighter parts often bring a strong commitment to justice, freedom, and even fun. They have immense dedication to protecting us from overwhelm. In a burdened system, sometimes our firefighters' behaviors are the only thing that grant us some release from both internal and external pressure.

Our exiles have strengths as well. They often hold truth and wisdom, even if they aren't fully updated to current situations. They hold needs for connection, safety, and authenticity. Sometimes our greatest strengths are the ones that our exiles embody but are afraid to show the world. If we were scolded for being too joyful or exuberant, perhaps those qualities got exiled. If we were told not to be so sensitive, perhaps even paying attention to our own feelings was something that got buried away. Helping our parts heal often leads us to the discovery of hidden skills, talents, and qualities that are deeply in alignment with the core of who we are.

The gifts that our parts have to offer don't go away when they become burdened. Unburdening frees parts so that they can contribute their skills and strengths to help us. They may develop new ways of using their skills that are more rewarding. Or they may have more sustainable boundaries. For example, a part that leads a person to overwork at their job may not be aware of the needs of that person's entire system, such as needs for sleep, rest, or social interaction. When this part is in awareness of and greater harmony with the rest of the system, it will be willing and able to "clock out" so that there can be more life balance.

We can have the goal of healing and unburdening while also maintaining respect and appreciation for what's working at this very moment. As with most things in life, it isn't all or nothing. Self leadership may be reflected in the capacity to have clarity about the nuances and contradictions that are inherent in the human experience.

The Interplay Between Individual and Community Resilience

Resilience is often talked about as if it's a trait, leading to assumptions about some people being more resilient than others. The idea that some people are just stronger than others and that all we need to do is "pull ourselves up by the bootstraps" is grounded in the myth of meritocracy. We all have strengths, but sometimes being blended makes it hard for us to see or access them. In our society, there's a tendency to attribute a person's success to their own individual strengths when, in fact, usually that person is being bolstered and supported by either the people who paved the way before them or the communities that have offered them resources along the way.

As queer and trans people, we've faced attacks on our right to exist. Our enduring strength and capacity to not only survive but also celebrate the magic of our queerness is evidence of our collective resilience. That collective resilience could be considered part of a collective Self energy that tells us to keep expanding into our truths even when the world tells us to contract, hide, or disappear.

We don't have to look too far to find examples of queer and trans community resilience. We know, as queer and trans people, that there's greater safety in numbers. When we connect with each other, whether it's at a queer in-person gathering or in an online space, we can access a greater sense of knowing that it's okay to be who we are. We've provided each other mutual aid and resources when governments and health care systems have denied our rights and needs. We've shared information with each other that has allowed us to navigate gatekeeping systems and access life-affirming, lifesaving health care. Much of the time we're exchanging information that helps us be informed about our bodies and health care needs beyond what medical providers are willing or able to offer us. We've provided physical and emotional care to each other after surgeries. We've created grassroots campaigns to advocate for the rights of incarcerated trans people. For decades, ball culture has been a safe haven for queer and trans communities, particularly those who are BIPOC. Despite recent attacks on drag performances, we continue to celebrate our right to creative expression. We know the meaning of "chosen family" because we've had to. The list goes on.

When we're able to tap into community resilience or Self energy, we're better able to access our individual or internal sense of resilience or Self energy. You might think of it as a constant feedback loop. Sometimes, especially when we're blended with parts that feel hopeless, we can only access the legacy burdens rather than legacy gifts or heirlooms. We may need help unblending so that we can recover our hope and perseverance.

The Guidance of Queer Ancestors and Trans-cestors

In some cultures, the emphasis on individualism draws our attention away from the greater context of our communities, including who and what in the past has allowed us to be where we are today. In many Indigenous cultures there's a reverence for and reliance on ancestral lineage. I refer to the term "lineage" broadly, as we know very well as queer and trans people that family isn't always defined by blood relations. Our practice of designating chosen family can extend to whom we call upon as ancestors. Many people I know consider queer and trans activists, artists, and movement makers as ancestors and trans-cestors. These include Audre Lorde, Leslie Feinberg, Marsha P. Johnson, James Baldwin, Frida Kahlo, Sylvia Rivera, and Bayard Rustin, among many others. We don't even know about many of the people who have laid the ground before us. I often use the term "ancestors known and unknown" because many powerful histories and movements have been erased by colonialism and white cisheteropatriarchy.

If at any time you feel that you need support beyond the Self energy that you can access in any given moment, you might draw on the energy or wisdom of someone you consider an ancestor. This practice is very personal, and sometimes you might not understand why you feel so connected to a person or historical figure. An ancestor might be from your own familial blood lineage, the culture that you grew up in, or other identities such as being a witch or an activist.

You may find different methods of connecting with these ancestors, such as having a photo or an object that represents them on an altar. If you're unsure of how to build a relationship with a guide or ancestor, or whether you even want one, that's okay. You can instead focus on certain qualities (e.g., wisdom, patience, perspective) that you imagine could help you and set an intention of opening yourself to receiving these qualities, possibly from an unknown ancestor, the universe, the realms of possibility, or whatever fits for you. In addition, the following meditation will guide you to connect to individual, community, and ancestral resilience and Self energy. An audio recording of this meditation can be found at http://www.newharbinger.com/55282.

Meditation: Queer and Trans Resilience

1. Identify a concrete object that has the quality of resilience. It could be anything: a plant, a stone, a book, an image of an ancestor or someone you admire, or even a favorite item of clothing that has withstood the test of time. Ideally, find something that you can touch or hold or place near you. If you can't find anything around you, you could even bring up a digital image on your phone.

2. Allow yourself to focus on the object you've chosen and bring to mind whatever qualities or strengths drew you to it. Notice your breath and what happens in your body as you continue to focus on these qualities.

3. Now turn your attention to whatever it was inside you that helped you identify this object. Go inside and see if you can connect with whatever that is. You may experience this as your Self energy, or you may experience this as a supportive part. Notice any images, sensations, feelings, or messages that are present.

4. Now see if you can set the intention to receive knowledge of or connection with your own **resilience** as a queer or trans person. Take a few breaths and notice

what comes up; it may be an image, a sensation, a message, an emotion, or something else. Can you sense guidance? Or a knowing that you aren't alone, that you don't have to do any of your healing work all by yourself? That you're part of a long history of ancestors with the power to support you? Or perhaps that you could even be a powerful ancestor for those to come? What positive qualities can you witness in yourself?

5. If you'd like, you can ask a guide or an ancestor to pass down to you any gifts or qualities that you're needing at this time.

6. If you find any of this difficult, you might also try a perspective shift: imagine yourself through the perspective of someone who's or has been supportive of you. When we're struggling, these people can mirror back to us all the ways that we are whole and worthy of care. Just know that even if you can't access or believe it in this moment, you're enough and you belong.

7. Take a few more moments to notice anything else that calls your attention.

8. When you're ready, open your eyes and write down anything that came up. This may include noting any of your positive qualities or ways that you've been respectful of yourself and your needs.

Pause and reflect: What came up for you during the meditation?

_____
_____
_____
_____
_____
_____

Summing It Up

In this chapter you've been guided to focus on strengths and resilience within your protective system. You've learned about the power of connecting to ancestral wisdom. If you'd like additional support in building resilience as a queer or trans person, *The Queer and Transgender Resilience Workbook: Skills for Navigating Sexual Orientation and Gender Expression* by Anneliese A. Singh is an excellent resource. For more information on inviting ancestral guides to support you in healing parts work, I highly recommend *Listening When Parts Speak: A Practical Guide to Healing with Internal Family Systems Therapy and Ancestor Wisdom* by Tamala Floyd. Having now connected with your own sense of resilience, in the next chapter we'll look at ways to apply this to your experience of navigating the world as a queer or trans person.

Chapter 12

Persisting in Self-Determination: Navigating the External World

As queer and trans people, navigating the external world presents certain challenges that not everyone in dominant culture has to face. In this chapter we'll discuss different spaces that can bring up feelings and reactions, which can then lead us to the parts that might benefit from more attention.

Being Perceived

Life would be a lot simpler if we didn't have to navigate other people's reactions to our identities, appearances, or relationships. On one hand, getting to a place where your parts are less concerned with what other people think can be incredibly freeing. On the other hand, being aware of other people's reactions to us can be crucial for our survival. Some of us have parts that have feelings about *not* being perceived as queer or trans. Being perceived can bring up a whole host of reactions from our parts:

"Look at me! See me!"

"Don't look at me! Pretend I'm not here!"

"I can't hold my partner's hand in public; it's too risky."

"What's that person thinking about me? Am I safe here? Will I be rejected?"

"I have to man/woman up to be taken seriously right now."

"This is too scary. Let's not leave the house anymore."

"I want to feel connected to other queer/trans people, but if they don't perceive my queerness/transness, then how am I supposed to do that? I feel invisible."

"Being perceived is a drag…literally!"

If you relate to these reactions within you, you're not alone. Your protectors may be working hard to manage how the external world views you, which can take a lot of energy for our systems. None of us has 100 percent control over how others experience us, but some of our parts may not know that.

We all have a relative amount of choice with regard to disclosing our queer and trans identities to others. Some of us have more choice than others. Some of us may blend in easily with the dominant culture or gender binary (some parts may want that, while others don't). Some of us may feel more obvious or *too* visible as queer or trans due to our appearances or ways of talking, moving, and interacting. Of course, there's no such thing as *too queer* or *too trans* because there's no such thing as being *too yourself*.

Disclosure: Balancing Safety and Authenticity

If you're navigating a choice about whether or not to come out or disclose your queer or trans identity to others, your parts may be polarized about what to do. On one hand, you might have parts that are excited about the possibility of living fully as your authentic self. On the other hand, you might have parts with valid concerns about what might happen or how others would treat you. These parts may be worried about physical or emotional safety, financial well-being, or access to resources.

On top of all of this, some parts may have judgments or opinions about either option. For example, a critical protector part may tell you you're not queer or trans enough, or that you're not being real, or that you're being dishonest if you don't come out. Queer and trans people hold the cultural burden of being accused of deception when in fact we live in a society that doesn't always honor or believe our truths. Another protector may believe that coming out is selfish and would hurt your loved ones. Whatever the case, it's important to notice parts with "shoulds" and, if possible, get curious about where these "shoulds" come from. Are these *your* values or beliefs? Or do they sound like someone else's voice? Or a message imposed by dominant (cisheterocentric) culture?

There's a lot more that could be explored about the choice to disclose your sexual orientation or gender identity to others. The emphasis here is on choice, and this isn't a one-time event. Many of us face choices every day, sometimes multiple times a day, about when and where to self-disclose. In this realm, Self energy might look like the capacity, from moment to moment, to be confident in our right to privacy and discernment.

Pause and reflect: How do your parts navigate being perceived? Take note of any of the parts that tend to lead when you're out in the world.

_____

_____

_____

Navigating Public Spaces

Dealing with larger systems can bring up a whole host of parts. It's almost impossible to move through life without encountering embedded assumptions about who exists in the world or how we're supposed to act based on our (perceived) gender or sexuality. This might be structural (e.g., binary gendered restrooms) or relational (e.g., other people trying to figure out gender or sexual orientation to inform how they interact with us). As queer and trans people, we know too well that navigating health care systems, employment, public assistance, housing, education, and the right to exist in or use public spaces isn't the same for everyone.

As stressful as these situations can be, we may feel like we can deal with them because we're just used to it. But our protective parts are likely working really hard to keep these challenges from breaking us. While we don't have complete control or power to change all of these systems today, we do have the choice to acknowledge the parts of us that are stressed or hurting from the day-to-day stress and tend to them by setting boundaries and advocating for ourselves.

Boundary Setting and Self-Advocacy

Dealing with external systems calls for us to be clear about our boundaries. A boundary helps us determine what is or isn't acceptable to us and to act in a self-respecting way when our limits are crossed. It's not about changing another person's behavior but rather what we're willing to subject ourselves to.

We may have parts that carry self-judgment shame for staying quiet or compliant. We'll want to remember that parts that keep us from setting boundaries have a protective intention. If it doesn't feel safe to assert our boundaries with others, we can still gain clarity about our boundaries with ourselves. If we're new to setting boundaries or don't fully believe that we have the right to set the boundary, we may not be so skillful. Our boundaries might show up rigidly or as walls (protectors).

Pause and reflect: Which parts are involved in setting boundaries (or not)? Check to see if any of your parts express fears or reactions to the idea of setting a boundary for yourself.

_____

_____

Doing parts work can also help us determine how we want to interact with larger systems and to respect our limits. When our parts are unburdened, we may also be able to access a different kind of energy for doing advocacy and activism work. We may feel more able to connect with community and collective resilience so that we can work toward the liberation of not just queer and trans people, but of all people who face marginalization.

This topic is quite broad, and I've intentionally left it general, as we each face different challenges in the world. The aim of the following worksheet is to identify trailheads or starting points for doing deeper parts work. A copy of this worksheet can be found at http://www.newharbinger.com/55282.

Worksheet: Identifying Trailheads in Response to the External World

In the spaces below, jot down any parts that you're aware of in different scenarios.

Situation	Parts I'm Aware Of
Dealing with immediate family members	
Dealing with extended family	
Dealing with bosses or coworkers	
Dealing with roommates	
Dealing with friendship circles	
Dealing with small queer community issues and overlaps	
Dealing with public spaces	
Dealing with capitalism	
Dealing with homophobic attitudes	
Dealing with anti-trans attitudes	
Dealing with the gender binary	
Dealing with health care providers	

What other parts am I aware of that come up regularly as I navigate the external world as a queer and/or trans person?

_____

_____

Now, having identified some of these parts, choose one to start working with. You may use the exercises in chapters 4 and 5 as a guide to connect with these parts. As you spend time with the different parts you identified, you may be able to navigate even the most challenging environments with more Self energy and peace of mind.

Building Friendships and Social Community

Having social community can be a key part of us feeling at home in our queer and/or trans identities, but access to this varies. Our geographical locations have a great influence on how easy it is to find queer events or social circles versus being more isolated. Additionally, meeting new friends is often informed by our earlier experiences and the burdens we carry. Growing up queer or trans can make it hard to trust or be vulnerable; sometimes our traumas (e.g., burdens) collide and make it hard to connect even when we want to.

Consider the parts of you that seek or avoid social connection. Which parts of you want to move toward other people, and which parts of you want to move away? For example, are there parts who want to meet new friends but fear being rejected? If so, this dynamic between your parts can make it harder to show up consistently in friendships or social interactions. On the surface, this might just appear as social anxiety when underneath there's a polarization among your parts.

For many queer people, chosen families are just as if not more important than families of origin because sometimes they are all we have. For example, in a lot of heterosexual marriages the couple unit is prioritized over all others that are "just" friendships. Of course, this can be true in some queer relationships as well, but as a whole we're more likely to value our friendship communities as much as if not more than romantic connections. In the following worksheet, you'll explore what it feels like for your system when there's more flow versus more challenge when connecting to others. A copy of this worksheet can be found at http://www.newharbinger.com/55282.

Worksheet: Building Community

Take a moment to bring up a time when you felt connected to another person. It could be anyone: a friend, partner, teacher, etc. Notice what it feels like in your body to think of this person. Notice if there are any qualities of Self energy that you can connect to. Take a moment to hear from any parts that have anything to share with you. Then thank these parts for allowing you to connect with them. Jot down anything you noticed.

_____

_____

_____

_____

Now think of a time when you were more *proactive* in seeking connection with others (whether as an individual or in a group). What strategies did your part(s) use in this instance?

_____

_____

_____

_____

If these parts are around, invite them to share with you. Invite them to tell you about their hopes as well as their fears or concerns. Thank these parts for sharing. Jot down anything you learned from these parts, including information about any other parts they're protecting.

_____

_____

_____

_____

Now think of a time when you were more *reactive* in a situation in which you wanted to connect with others (whether as an individual or in a group). What strategies did your part(s) use in this instance?

_____
_____
_____
_____

If these parts are around, invite them to share with you. Invite them to tell you about their hopes as well as their fears or concerns. Thank these parts for sharing. Jot down anything you learned from these parts, including information about any other parts they're protecting.

_____
_____
_____
_____

If you had a hard time connecting with any parts, or if you're most connected with parts that have strong objections to building social connection, take a moment to acknowledge these parts. Invite them to share anything they would like you to know, including whatever fears or concerns they have about what would happen if they allowed you to feel connected to other people. Thank them for sharing with you. Jot down anything you noticed.

_____
_____
_____
_____

Now take a moment to reflect on the entire exercise. Did you learn anything new? Which parts (protectors) might you want to spend more time with?

_____

_____

_____

_____

Summing It Up

In this chapter we explored some of the issues that might arise when navigating external systems as a queer and/or trans person. We looked more closely at what boundary work might look like from a parts perspective and were invited to identify some potential trailheads to work with. We also looked at cultivating social connection to help us remember we don't have to face life challenges alone. The next chapter will guide you in exploring your relationship to your mental health.

Chapter 13

Mental Health Support

Mental health can be influenced by many factors: cultural, familial, individual, spiritual, and relational. A comprehensive review of all aspects of mental health is outside the scope of this book. However, in this chapter we'll explore how a parts framework can relate to how we might view seeking mental health and emotional support.

Applying a Parts Lens to Common Mental Health Concerns

Western psychiatric disorders or classifications of symptoms typically don't capture the full picture of what's happening for a person's emotional or mental health. Parts work can be used in conjunction with psychiatry or mental health disorders frameworks, but in IFS we have a more nuanced view of what's happening. Because psychiatric diagnoses vary considerably across time and culture, they can be immensely helpful for some people but for others completely miss the mark. Some diagnoses are overly dependent on the concept of functioning, which is ableist and can serve to both underdiagnose or overdiagnose mental disorders.

In IFS we don't simply point at behaviors and call them dysfunctional or disordered. We know that parts are doing what they do for a reason, and in some ways they're quite "functional" in our systems. We understand that parts that are causing problems for us are also trying to help us cope or survive deeper pain.

So how do we understand mental health diagnoses or disorders from a parts perspective? There isn't just a one-to-one relationship between diagnoses and parts. In other words, there isn't just one depression part or one ADHD part or one OCD part. Each diagnosis might be comprised of several parts in a dynamic relationship with each other.

Remember that everyone's experience and internal system is different as you reflect on your own experiences. Here are just a few examples of how mental health concerns might be expressed in an internal system:

- Anxiety can show up in the form of a protector or group of protectors who are protecting an exile from getting hurt (again). This may include parts that are hypervigilant, parts that are expressed as body sensations, parts that come up with worst-case scenarios, or parts that avoid social situations (to avoid rejection, to maintain safety).

- Depression can involve not just parts that are sad, but also parts that shut down emotions, parts that are self-critical, or parts that isolate, parts that are easily irritable, or parts that consider suicide as a way to escape pain.

- Gender dysphoria can have parts similar to those in anxiety or depression, but it may also include parts focused on one's body/gender, parts that feel pain when being perceived or treated in ways that don't align with one's identity, or parts that develop ways of coping with other people's responses.

- Obsessive-compulsive disorder (OCD) can involve parts that ruminate or obsess, parts that embody compulsive behaviors or rituals, or even parts that are fixated on morality.

- ADHD may include parts that hyperfocus, parts that are easily distracted, or related parts that mask in order to fit in or avoid rejection.

- Addictions may include parts that use reactive or extreme strategies to numb pain, parts that try to control or moderate the behavior, and other parts that have shame or judgment.

- Disordered eating exists on a wide spectrum and may include parts that obsess about appearance, parts that try to control, and other parts that seem "out of control."

Keep in mind that although the parts involved in mental health symptoms or diagnoses can cause us problems, they all have a positive protective function; these parts may not be aware that their strategies don't work or aren't adequate long-term solutions. The following meditation will guide you in getting to know a "symptom" (in quotations because this is a medicalized term); if other words are a better fit for you, please use those (e.g., concern, protective strategy, survival mechanism, stress signal). An audio recording of this meditation can be found at http://www.newharbinger.com/55282.

Meditation: Getting to Know a "Symptom"

1. Begin by finding a comfortable position. Take a few deep breaths.

2. Think of one specific symptom or experience that impacts your mental health or gets in the way of feeling emotionally balanced. Other connected experiences may show up, but for now just start with one. It might be avoiding feelings, using substances to deal with stress, engaging in food restriction, or something else.

3. Check to see how open you are to connecting with the part that expresses itself through this behavior or experience. If you feel open, continue on. If you notice any criticism or judgment, ask that part to step back so you can have this conversation.

4. When you're ready, see if the part that uses this symptom to protect you is around. Notice how you're experiencing it. Find where it is in or around your body. Notice any other sensations, emotions, or images.

5. Check again to see how you're feeling toward this part. If you feel open to hearing from it, let it know that you'd like to connect and understand it better. See if it's open to a conversation. If so, ask the part the following questions:

    - What's your role?
    - What's your intention for me? How are you trying to help?
    - Do you like being in this role, or is there something you'd rather do?
    - How did you come to be in this role?
    - What are you worried would happen if you didn't do what you do?
    - What other parts need attention or help?

- If there were a way for me to help these parts so that you didn't have to work so hard at protecting me in this way, would you be interested in that? What would need to happen for you to allow me to step in and take care of you and these other parts?

6. After hearing from these parts, thank them. Now take some time to jot down anything that came up.

Pause and reflect: What came up for you during the meditation? Has anything shifted in how you view the "symptom" you focused on?

_____

_____

_____

_____

Compassion versus Shame: Do Parts Allow You to Ask for Help?

Even though IFS work is based on the premise that we, in being present for our parts, have the power to heal, that doesn't mean that we must only rely on ourselves or that we shouldn't ask for outside help. Freeing ourselves (our exiles) of burdens can open up many possibilities for accessing helpful resources or collective/community power.

There are a number of burdens or internalized negative beliefs that our parts may carry about asking for help, such as:

- It's not okay or safe to ask for help.

- Asking for help means that I'm weak. I should be able to figure this out on my own.

- People will judge me if they know I suffer from a mental health disorder.

- People in my family or culture don't go to therapy.

- My friends won't understand.

- What's the point? No one is going to help me anyway. It's better to not ask rather than risk disappointment.

The following meditation will help you reflect on what parts may be involved in how easy it is for you to ask for help. An audio recording of this meditation can be found at http://www.newharbinger.com/55282.

Meditation: Asking for Help

1. Get into a comfortable position. Start to notice your breath.

2. Consider the idea of reaching out for help with something you're struggling with—regardless of how big or small it is—and then notice what shows up in your awareness. Just notice any feelings, any reactions, any parts that have objections to this idea.

3. Without trying to respond or fix these parts, just notice. Notice their energy. Notice if there are any sensations that are coming up in your body. Notice however these parts are trying to express themselves. As much as possible, offer space to these parts to be witnessed by you.

4. If you're having a hard time hearing from any of these parts, take a moment to hear from any other parts that are having reactions or judgments. Ask those parts if they would be willing to soften back so that you can hear from the ones that are having a hard time with the idea of asking for help.

5. After you've heard from these parts, ask them what they're worried would happen to you if you did ask for help. If what they're sharing makes sense to you given what you've learned or experienced, let your parts know that.

6. If it's clear that your parts are protecting other vulnerable parts, ask if they would let you build some connection with the vulnerable parts. Let them know that with your help, with Self energy, these parts may not need to work as hard to protect you or the vulnerable parts.

7. Take another few moments to connect in with any of the parts that have anything else to share with you. When you're at a stopping point, jot down anything that you want to remember or continue to work with.

Pause and reflect: What came up for you during the meditation?

___

___

___

___

___

Summing It Up

In this chapter you've gone over a brief review of different ways that parts can show up as mental health concerns. You've learned that mental health problems or diagnoses aren't typically just one part, but rather a dynamic set of interactions among multiple parts. You've gotten a chance to get to know one aspect of a part that affects your mental health, as well as gain a better understanding of what it means for you to ask for help. In the next and final chapter, we'll explore how to consistently integrate parts work in everyday life.

Chapter 14

Establishing a Continuous Practice

We've covered quite a bit of ground, yet this is only a beginning. By this time, you probably have some experience connecting with different parts of you. However, because we're dynamic and complex beings who are always responding to different circumstances and environments, we'll continue to get blended and (hopefully) engage our parts consciously so that they're willing to unblend and allow Self leadership. And, in knowing our systems better, we'll be able to recognize when we're blended and return to being Self-led with more ease. In this chapter we'll discuss how to establish a continuous, consistent practice of building deeper relationships with your internal system.

Tips for Moving Forward

1. Balance expectations: remember this is a new practice. Embracing the concept of having a "beginner's mind" might allow you to enjoy the process of learning and applying IFS. If you feel even 1 percent more access to Self energy or connection with your parts, consider that a win.

2. Notice what allows you to experience the most Self energy and *move toward that* as much as possible. You can ask yourself: Does this person, place, activity, or thing help me access Self energy?

3. When doing parts work, continue to ask yourself, "Who's leading?" Be aware that perfectionistic parts or those who have a strong urgency to heal your system will likely slow down or get in the way of your process.

4. Find some way to keep track of your parts. Parts that need your attention will continue to show up, so tracking which parts you've met or spent time with can help you maintain perspective or feel like you have a road map for your internal work.

5. Consider approaching this practice as a "habit." Habits can only be solidified with consistency, so try to find some way to talk with your parts every day for at least thirty days. The easiest way to do this is to be realistic about what's possible. Set aside ten minutes every day. If that's too much, set aside five minutes. If that's too much, set aside two minutes.

6. Try your best to remember that multiplicity applies to everyone. This can really create ease in how you go about your days or how you relate to others. There's something qualitatively different about thinking "I can't stand that person!" and "Part of me can't stand that part of that person."

7. Use this book as a resource and reference. You may have to read and do the exercises several times before you feel a sense of comfort or confidence.

8. Find a buddy who's also interested in IFS and, after both working through this workbook (either on your own or together), set up a regular accountability check-in. Daily or weekly will be most supportive of your continued practice.

9. Find an experienced IFS therapist or practitioner (the IFS Institute has a directory). If you're doing one-on-one healing work with a healer or professional, you may want to share this workbook with them.

10. Find other resources! There are books, podcasts, support groups, and online communities of people practicing IFS.

The following worksheet can help you come up with a plan for continuing to practice and integrate parts so that you can feel more connected to yourself. A copy of this worksheet can be found at http://www.newharbinger.com/55282.

Worksheet: Creating Your Own Plan

Are there any concepts, exercises, or practices that felt particularly impactful for me or my internal system? How might I continue to apply those to my life?

_____

_____

_____

What are three things that I can do to set myself up for continuing to practice IFS moving forward?

    a. _____

    b. _____

    c. _____

Are there any obstacles I'm anticipating, and what might I need to do to address them?

_____

_____

Who can I talk to about this practice? Would it be helpful to share the parts or multiplicity perspective with a friend, partner, or family member?

_____

_____

What additional resources do I want to seek out that are in line with this practice?

_____

_____

May the Self Be with You…and All of Us

Thank you for taking the time to invest in a better relationship with yourself and your internal system of parts. I hope that the IFS approach continues to support you in living in greater and greater alignment with your magical queer and/or trans self and your communities. I truly believe that when we, as queer and/or trans individuals, engage in liberatory practices within ourselves, we have a much better chance of moving toward collective liberation for all. I'm rooting for you and for all of us!

Acknowledgments

I've been immensely fortunate to have learned from many brilliant and talented IFS practitioners. I want to express deep gratitude to: Nic Wildes for being my co-instructor and thought partner in this journey of building our Queer and Trans Internal Family Systems (QTIFS) community; Tamala Floyd, Kim Paulus, Pamela Krause, Chris Burris, Crystal Jones, Cathy Curtis, Nancy Sowell, and many others for your wisdom and mentorship; and Richard Schwartz and all who have contributed to sharing this powerful and evolving model with the world. I also thank our QTIFS staff and participants for your dedication.

Thank you to the people at New Harbinger who have guided the process of bringing this book to life: Elizabeth Hollis Hansen, Madison Davis, and Karen Levy.

Thank you to my clients for trusting me to witness them coming back to themselves.

Thank you to my family, friends, therapists, and recovery community for supporting and listening to me and my hyperverbal parts!

Last but not least, thank you to my wisest teachers and embodiments of "all pawts welcome," my sweet pups BonBon Jovi, Zelda Sesame Mochi Chang, and Theo Chang, the pugasus in the sky.

Glossary

Backlash: A part or internal system's reaction when healing or an internal shift occurs too quickly; a part may have an overwhelming response or attempt to shift the system back to the status quo or previous equilibrium.

Blended: Parts-led rather than Self-led; a state in which a person has less access to Self energy and is being led by a part or parts. Being blended exists on a continuum.

Body Liberation: A broad movement away from the structures, policies, behaviors, beliefs, and ways of relating that are based on a "right" or "normal" way to have a body as defined within a given group or culture.

Body Trust: A universal innate capacity to *listen* to the wisdom of one's body and act accordingly. Could be seen as an expression of Self leadership.

Burdens: A painful belief, thought, feeling, behavior, or sensation that an exile carries about itself or the person whose system it exists in. Burdens are acquired through direct experience or inherited from family, ancestral lineages, or the larger culture.

Centered Part: A part that's the agreed-upon focus of an IFS conversation with Self; typically referred to as a "target part."

Code-Switching: The skill of presenting oneself differently depending on the social context; often discussed in the context of being a person with a marginalized identity navigating dominant culture.

Cultural Burden: A painful belief, thought, feeling, behavior, or sensation held by an entire group of people, often the result of past or present marginalization.

Exiles or Exiled Parts: Parts that hold painful memories, negative beliefs, traumatic experiences, or wounds; they're often young or vulnerable and developmentally stuck or frozen in time.

Firefighter or Firefighter Parts: A protector part that's reactive, impulsive, or intent on extinguishing or numbing overwhelming feelings.

Legacy Burden: A painful belief, thought, feeling, behavior, or sensation that was passed down through prior generations and is typically not the result of a person's own lived experience.

Manager or Manager Parts: A protector part that's proactive, focused on controlling the external world or other parts, or trying to prevent negative experiences.

Masking: The process by which (consciously or unconsciously) a person behaves in ways that conceal their true nature or ways of being in order to fit in with the external world or prevent rejection; often discussed in the context of neurodivergence.

Multiplicity: Being made up of distinct parts or subpersonalities, each with their own experiences, perspectives, or roles. Multiplicity is a universal human experience.

Parts-Led: A state in which a person is guided by a part or parts; blended.

Polarization: A dynamic in which two or more parts are in conflict with each other or hold opposing goals or strategies within a person's internal system. Polarizations may lead to indecision or overwhelm.

Protector or Protector Parts: Parts that try to shield a person from experiencing hurt or harm. The two types of protectors, managers and firefighters, both seek to protect exiles.

Retrieval: The process of helping a part leave a scene, situation, or moment in time that it has been developmentally stuck in. The part may be given the choice of where to go (e.g., a pleasant place, the present day with Self).

Self or Self Energy: The core or essence of a person; an innate wisdom, knowing, or consciousness that can witness and care for all parts.

Self-Led: A state in which a person is guided by Self rather than their parts; less blended or unblended.

Self-to-Part Relationship: The connection between Self and various parts. Self-to-part relationship is considered the vehicle for healing in IFS.

Titration: An approach that involves doing healing or trauma recovery work in small, tolerable amounts so as not to overwhelm the system.

Trailhead: A starting point or entry into a life problem or issue that's raising concern.

Updating: The process of helping a part become aware of present-day situations or realities, including who Self is, what year it is, what the person's life looks like now, and past circumstances no longer being true.

References and Further Reading

binaohan, b. 2014. *decolonizing trans/gender 101*. https://publishbiyuti.org/decolonizing transgender101.

Catanzaro, Jeanne. 2024. *Unburdened Eating: Healing Your Relationships with Food and Your Body Using an Internal Family Systems (IFS) Approach*. Eau Claire, WI: Bridge City Books.

Coleman, Eli., A. E. Radix, W. P. Bouman, G. R. Brown, A. L. De Vries, M. B. Deutsch, et al. 2022. "Standards of Care for the Health of Transgender and Gender Diverse People, Version 8." *International Journal of Transgender Health* 23 (sup1): S1–S259.

Floyd, Tamala. 2024. *Listening When Parts Speak: A Practical Guide to Healing with Internal Family Systems Therapy and Ancestor Wisdom*. New York: Hay House LLC.

Glass, Michelle. 2016. *Daily Parts Meditation Practice: A Journey of Embodied Integration for Clients and Therapists*. Listen3r.

Johnston, Alo. 2023. *Am I Trans Enough?: How to Overcome Your Doubts and Find Your Authentic Self*. Philadelphia: Jessica Kingsley Publishers.

Kinavey, Hilary, and D. Sturtevant. 2022. *Reclaiming Body Trust: A Path to Healing & Liberation*. New York: Penguin.

Kinsey, Dalia. 2022. *Decolonizing Wellness: A QTBIPOC-Centered Guide to Escape the Diet Trap, Heal Your Self-Image, and Achieve Body Liberation*. Dallas: BenBella Books.

Lehrner, Amy, and R. Yehuda. 2018. "Trauma Across Generations and Paths to Adaptation and Resilience." *Psychological Trauma: Theory, Research, Practice, and Policy* 10 (1): 22.

Rasmussen, Sofie M., M. K. Dalgaard, M. Roloff, M. Pinholt, C. Skrubbeltrang, L. Clausen, and G. Kjaersdam Telléus. 2023. "Eating Disorder Symptomatology Among Transgender Individuals: A Systematic Review and Meta-Analysis." *Journal of Eating Disorders* 11 (1): 84.

Schwartz, Richard C., and A. Morissette. 2021. *No Bad Parts*. Boulder, CO: Sounds True.

Singh, Anneliese A. 2018. *The Queer and Transgender Resilience Workbook: Skills for Navigating Sexual Orientation and Gender Expression*. Oakland, CA: New Harbinger Publications.

Strings, Sabrina. 2019. *Fearing the Black Body: The Racial Origins of Fat Phobia*. New York: New York University Press.

Sykes, Cece, M. Sweezy, and R. C. Schwartz. 2023. *Internal Family Systems: Therapy for Addictions*. Eu Claire, WI: PESI.

Sand C. Chang, PhD, (they/them) is a Chinese American nonbinary, queer, and neurodivergent clinical psychologist, consultant, and somatic psychotherapist based in California. They are a Level 3 certified internal family systems (IFS) therapist, trainer at the IFS Institute, certified Body Trust provider, and cofounder of Queer and Trans Internal Family Systems (QTIFS). Learn more about Sand at www.sandchang.com.

Real change *is* possible

For more than fifty years, New Harbinger has published proven-effective self-help books and pioneering workbooks to help readers of all ages and backgrounds improve mental health and well-being, and achieve lasting personal growth. In addition, our spirituality books offer profound guidance for deepening awareness and cultivating healing, self-discovery, and fulfillment.

Founded by psychologist Matthew McKay and Patrick Fanning, New Harbinger is proud to be an independent, employee-owned company. Our books reflect our core values of integrity, innovation, commitment, sustainability, compassion, and trust. Written by leaders in the field and recommended by therapists worldwide, New Harbinger books are practical, accessible, and provide real tools for real change.

MORE BOOKS from NEW HARBINGER PUBLICATIONS

THE QUEER AND TRANSGENDER RESILIENCE WORKBOOK

Skills for Navigating Sexual Orientation and Gender Expression

978-1626259461 / US $25.95

THE MENTAL HEALTH GUIDE FOR CIS AND TRANS QUEER GUYS

Skills to Cope and Thrive as Your Authentic Self

978-1648485039 / US $19.95

I'M NOT OKAY AND THAT'S OKAY

Mental Health Microskills to Deal with Life's Inevitable Struggles

978-1648481758 / US $18.95

CUTTING TIES WITH YOUR PARENTS

A Workbook to Help Adult Children Make Peace with Their Decision, Heal Emotional Wounds, and Move Forward with Their Lives

978-1648483905 / US $25.95

THE POLYVAGAL SOLUTION

Vagus Nerve-Calming Practices to Soothe Stress, Ease Emotional Overwhelm, and Build Resilience

978-1648484124 / US $19.95

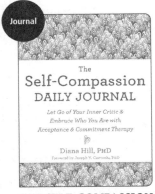

THE SELF-COMPASSION DAILY JOURNAL

Let Go of Your Inner Critic and Embrace Who You Are with Acceptance and Commitment Therapy

978-1648482496 / US $18.95

newharbingerpublications

1-800-748-6273 / newharbinger.com

(VISA, MC, AMEX / prices subject to change without notice)

Follow Us

Don't miss out on new books from New Harbinger.
Subscribe to our email list at **newharbinger.com/subscribe**

Did you know there are **free tools** you can download for this book?

Free tools are things like **worksheets**, **guided meditation exercises**, and **more** that will help you get the most out of your book.

You can download free tools for this book—whether you bought or borrowed it, in any format, from any source—from the New Harbinger website. All you need is a NewHarbinger.com account. Just use the URL provided in this book to view the free tools that are available for it. Then, click on the "download" button for the free tool you want, and follow the prompts that appear to log in to your NewHarbinger.com account and download the material.

You can also save the free tools for this book to your **Free Tools Library** so you can access them again anytime, just by logging in to your account! Just look for this button on the book's free tools page.

+ Save this to my free tools library

If you need help accessing or downloading free tools, visit **newharbinger.com/faq** or contact us at **customerservice@newharbinger.com**.